UNDERSTANDING EPIDEMIOLOGY

Concepts, Skills, and Applications

Revised First Edition

By Laura Wheeler Poms, PhD MPH
George Mason University
and Rebecca Smullin Dawson, PhD MPH
Allegheny College

cognella®
academic publishing

Bassim Hamadeh, CEO and Publisher

Michael Simpson, Vice President of Acquisitions

Jamie Giganti, Managing Editor

Miguel Macias, Graphic Designer

Amy Stone, Acquisitions Editor

Mirasol Enriquez, Project Editor

Alexa Lucido, Licensing Associate

Sarah Wheeler, Interior Design

First published in the United States of America in 2015 by Cognella, Inc.

Trademark Notice: Product or corporate names may be trademarks or registered trademarks, and are used only for identification and explanation without intent to infringe.

Cover images:
  Copyright © 2010 Depositphotos Inc./dimmushka.
  Copyright © Shutterstock Images LLC/Jezper.
  Copyright © 2012 Depositphotos Inc./Kuzmafoto.
  Copyright © 2012 Depositphotos Inc./songbird839.
  Copyright © 2012 Depositphotos Inc./CandyBoxImages.

Interior image copyright © 2012 Depositphotos Inc./sergio77.

Printed in the United States of America

ISBN: 978-1-63487-002-3 (pbk) / 978-1-63487-003-0 (br) / 978-1-63487-282-9 (pf)

www.cognella.com   800-200-3908

DEDICATION

To my parents, Chuck & Gini, who taught me the value of hard work and have never stopped encouraging me.

—RSD

To the entire Wheeler and Poms clans who, while they may have often doubted my sanity, always encouraged me to pursue my passion.

—LWP

CONTENTS

PREFACE

We decided to write *Understanding Epidemiology* because we are passionate about teaching undergraduate students the science of public health. We are committed to introducing epidemiological concepts and theories to our students in a way that is understandable and can be applied in real-world settings. Our hope is that others will use this book to educate undergraduate students interested in studying the patterns, distributions and etiology of disease as a means of improving public health.

The primary purpose of this book is to teach undergraduates the science of epidemiology. We have designed the book to be interdisciplinary. It covers epidemiological study designs and calculations for measuring disease. Additionally, we discuss the history of epidemiology, how epidemiology is used in public health practice, and the ethical requirements of conducting research with human participants. In addition to introducing students to the science, the book also teaches hypothesis generation and testing and quantitative reasoning skills; as well as skills to interpret and apply the findings of epidemiological studies.

This textbook was written so that students do not need to have a working knowledge of biostatistics to understand and apply the epidemiological concepts. Additionally, we recognize that not all students interested in studying epidemiology have strong mathematical skills or a fondness for math. We, therefore, have developed the exercises and examples throughout this book so that those unsure of their mathematical skills can understand the concepts presented.

We expect students using this textbook will be able to: 1) generate hypotheses about diseases and potential risk factors; 2) design epidemiological studies to test these hypotheses; 3) discuss the strengths and limitations of each epidemiological study design; 4) calculate appropriate measures of association between diseases and risk factors; 5) critically read both the epidemiological and public health literature; and 6) use data from epidemiological studies to evaluate the use and effectiveness of public health programs and policies.

Our book provides many examples of key concepts and opportunities to practice important epidemiological skills. These activities are designed to encourage students to think critically about public health problems and prepare them for real-world applications of the material.

ACKNOWLEDGMENTS

We gratefully acknowledge the contributions of a number of people to this book.

We thank the mentors and colleagues who have contributed to our understanding of epidemiology and encouraged us as educators.

I, Becky, specifically want to thank Eric Pallant at Allegheny College. The beginning of my public health career begins with you and a grassroots campaign to eliminate the use of PVC plastics in hospitals. Thank you for your guidance and unending support throughout my undergraduate career; encouraging me to complete my PhD; and welcoming and mentoring me as a faculty member. It is an honor to call you a friend and colleague. And to Drs. Laura Hungerford and Anthony Harris at the University of Maryland, Baltimore, much of what I learned about epidemiology and how to effectively teach it I learned in your class and through working as your teaching assistant. You inspired me to become an educator. Thank you for your encouragement, support, and guidance throughout my doctoral program.

I, Laura, wish to thank Dr. Paige Wolf, at George Mason University's School of Business. I learned much of what I know about teaching a diverse student body from your excellent advice, support and coaching. I greatly appreciate my dissertation advisor, Dr. Lois Tetrick, also at George Mason University, who taught me the value of persistence, perseverance and patience. You encouraged me to explore methodologies that would complement my work in occupational health psychology and consequently, I found epidemiology. Thank you so much for all that you did to get me where I am today!

Special thanks to Allegheny College students Erica Bryson, for her work editing the text, compiling the glossary, testing the exercises, developing answer keys, and providing valuable feedback; and Elizabeth Schafer, for providing valuable assistance in locating references used throughout the book. And to Frances Cole, who edited the text and provided valuable feedback. Our appreciation is also extended to the entire team at Cognella, who worked patiently with us to make this book a reality.

We are grateful for the supportive environments at our home institutions, Allegheny College and George Mason University, which provided us with the encouragement to write this book; and to all of the students in our epidemiology classes, who have challenged us to be better educators and inspired us to write this book.

And last, but certainly not least, we want to thank our families for their love and support.

I, Becky, struggle to find the right words to thank my husband, David, who has encouraged and supported me as I simultaneously wrote this book and I chased after my dream job. You have been my biggest cheerleader and a faithful friend and partner. Thank you for believing in me when I didn't believe in myself and for loving me. I am honored to call you my husband and friend. And to our children, Kennedy and Colin, I love both of you to the moon and back. It is my prayer that you will each find your calling in life and pursue it passionately.

I, Laura, greatly appreciate the patience, understanding and ability to tolerate carry-out dinners displayed by my husband Keith and daughters Allison and Kate, during this long process. I don't think my girls would know what to do if Momma wasn't writing something. My husband will now see more of me and hopefully that will be a good thing. Thanks also go to my parents, Bettie and Larry Wheeler, who have always said I should write a book. While I am not sure this is exactly what they had in mind, I know they are proud of me. It has been a long and winding road to get to this place, but I would not have changed a step along the way.

1 WHAT IS EPIDEMIOLOGY?

WHAT IS EPIDEMIOLOGY?

Public Health and Epidemiology

There is no single definition of health. The standard definition is the absence of disease, but in fact, there is much more to health than this. More than 65 years ago, the World Health Organization (WHO) defined health as "physical, mental and social well-being" and not merely the absence of disease. When this definition is applied to society at large, it becomes public health: an organized effort aimed at reducing **mortality** (death) and **morbidity** (disease or injury) while improving the health of populations. Thus, the focus of epidemiology is on the occurrence of health and disease in the population. A **population** is all the inhabitants of a given area considered together. This population approach contrasts with clinical medicine's concern with the individual. Epidemiology is sometimes called population medicine.

Epidemiologists study the distributions and causes of health and disease, injury, disability, and death in populations, as opposed to medicine, which focuses on individual health. The results of epidemiological studies are applied to the prevention and control of disease and the promotion of health. Epidemiological studies provide information that informs public health policy and planning, helping policy makers use scarce health resources to establish health programs with appropriate priorities. While the main goal of epidemiological studies is to address population health issues, these studies also provide individuals with information that can promote better health. To summarize, epidemiology aids with health promotion, the alleviation of adverse health outcomes, and the prevention of disease.

The Role of Epidemiology and Epidemiologists

Epidemiologists participate in all aspects of public health, from evaluating the safety of the food supply to discerning likely causes of common chronic disease like cancer or diabetes. They may work in global health, focusing on eradicating the last cases of polio or on some other equally important infectious disease. Some epidemiologists monitor diseases over time to look for changes in risk factors or differences in outcomes, identifying populations at greatest risk for a particular disease. Epidemiologists evaluate the effectiveness of prevention and treatment programs and are often involved in clinical trials evaluating the efficacy of new drugs. Often, epidemiologists are seen on television or are quoted in news articles as they discuss public health issues such as pandemic flu, emergency preparedness, or environmental concerns. Epidemiological information is used in health program planning and decision making, helping to determine where limited funds and resources can best be spent to improve the health of populations.

Epidemiology is a scientific discipline and uses research methods similar to those used in the basic sciences like biology, chemistry, and physics. But it is also an interdisciplinary science because epidemiology studies the human response to exposure and disease,

Figure 1.1 An example of a health education campaign, this 1964 poster featured the Center for Disease Control's (CDC) national symbol of public health "Wellbee," who reminded the public to "be well, be clean and WASH YOUR HANDS."

using information from many fields, including: mathematics and biostatistics; behavioral and social sciences; demography and geography; and law.

Epidemiology is also an observational science, capitalizing upon naturally occurring situations in order to study the occurrence of disease. Individuals who have experienced a unique exposure are often followed over time to see the health outcomes of that exposure. For example, residents of the Tohoku area of Japan experienced a magnitude 9.0 earthquake followed by a tsunami on March 11, 2011. This was the fourth largest earthquake in the world and the largest in Japan since modern seismic recordings began in 1900.

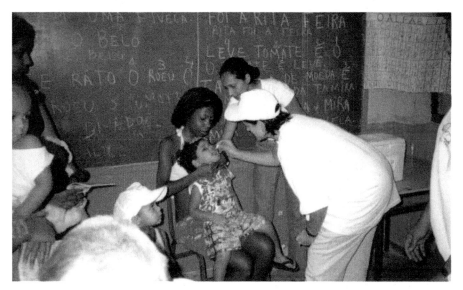

Figure 1.2 A National Immunization Day in Brazil. Epidemiologists work to organize the event in addition to tracking each child receiving a vaccination.

The tsunami was the most deadly since the 2004 Sumatra earthquake and tsunami. More than 15,000 people were killed, and hundreds of thousands of residents were exposed to nuclear radiation when the Fukushima Daiichi nuclear plant was badly damaged. These individuals will be closely monitored over a long period for any health effects that may occur years after the exposure. While they are monitored, they also receive the appropriate preventive medication to counteract any known possible negative effects from the exposure.

Epidemiological findings contribute to the prevention and control of disease, injury, disability, and death by providing information leading to informed public health policy and planning, as well as individual health decision making. In other words, the findings of epidemiologists must influence public policy because public policy is a major determinant

Figure 1.3 Tsunami damage in Kesennuma, Miyagi, Japan. Severe damage to the area's infrastructure, including a nuclear power plant, may cause health problems for many years to come. Epidemiologists will monitor the health of these individuals.

of population health. By measuring health and disease, epidemiology aids with health promotion and health education. Epidemiology contributes to public health by analyzing the evidence from community and laboratory-based studies, ensuring that important health problems are properly addressed.

2 A BRIEF HISTORY OF EPIDEMIOLOGY

The roots of epidemiology can be found all the way back to ancient Greece. **Hippocrates**, revered as the father of medicine, is credited with departing from superstitious reasons and supernatural explanations for disease outbreaks. In his *On Airs, Waters, and Places*, written around 400 BCE, he hypothesized disease might be associated with environmental factors such as air and water quality. He also suggested that individual behavior played a role in human health, often asking patients about their food and drink habits. Hippocrates was the first to identify the association of malaria and yellow fever with swampy regions, attributing the diseases to the bad air found in these areas. In fact, for thousands of years, it was not known that the mosquito was responsible for these diseases. It was not until 1900, when Walter Reed, a U.S. Army physician working in the tropics, made the epidemiological connection among the mosquito, the swampy environment and yellow fever.

Fast-forwarding several centuries from the time of Hippocrates, we begin to see epidemiology emerge. Auroleus Phillipus Theostratus Bombastus von Hohenheim, known as **Paracelsus** (1493–1541), is considered one of the founders of toxicology, the branch of science concerned with the study of chemicals and their effects on the human body. He established the concept of the dose-response relationship, which states that the effect of a toxic substance on the body is directly related to the strength of its dose. Establishing a dose-response relationship is one of the ways modern epidemiologists determine causal relations between exposures and outcomes. Paracelsus also developed the notion of target-organ specificity of chemicals, finding that toxic substances do not affect all organs to the same extent. For example, one chemical may adversely affect the liver, while another may poison the kidneys.

Moving a little further ahead through time to the 17th century, **John Graunt** (1620–1674) is credited with being the first to notice predictable variations in births and deaths. Working in London, he systematically recorded age, sex, cause of death, and where and when death occurred. Graunt divided deaths into two types of causes: acute, in which the person was struck suddenly (e.g., cholera), and chronic, in which the disease lasted over a long period time (e.g., emphysema). Using these vital statistics, he developed and calculated life tables. Graunt's work laid the foundation for epidemiology and demography. Life tables are still used in modern times by epidemiologists to analyze how long a patient with a particular condition is likely to survive.

Thomas Sydenham (1624–1689) was quite the maverick of his time. This English doctor believed in an empirical approach to medicine and emphasized that observation should drive the study of disease. He observed the natural course of a disease, uninfluenced by the traditional medical theories of the day. Sydenham described and distinguished different diseases, including some psychological conditions. He also suggested treatments such as exercise, fresh air, and a healthy diet, considered conventional wisdom today, but rejected by other physicians at that time. Sydenham was often persecuted by other doctors for his unconventional approach to health; however, his patients (and more recently trained physicians) appreciated his uncommon remedies.

Bernardino Ramazzini (1633–1714), an Italian physician, is regarded as the founder of the field of occupational medicine. He authored *De Morbis Artificum Diatriba* (Diseases of Workers), published in 1700. This book considered the health hazards found in 52 different occupations. He discussed chemicals, dust, metals, repetitive or violent motions, unnatural postures, and other hazards experienced by workers that could lead to illness or chronic impairment. For example, he identified that workers in certain occupations such as glassworking and blacksmithing were prone to eye injuries, and that bakers and those in the clothing industry were likely to develop respiratory diseases from breathing in fine dust particles.

Ramazzini was very interested in the prevention of occupationally related illnesses. When he found that workers who cleaned cesspools and privies often went blind from working in such toxic environments, he suggested these workers fasten transparent "bladders" over their eyes to protect them. He also recommended changes, including adequate ventilation and protection from extreme temperatures, in the working environment to help workers. Ramazzini's seminal work led to the more modern perspective of considering occupation and one's work environment when assessing an individual's overall health.

Percivall Pott (1714–1788), in addition to being an eminent doctor and surgical writer, is thought to be the first person to describe an occupational cause of cancer. In 1775, he found an association between exposure to soot and a high incidence of scrotal cancer in chimney sweeps, a malady specific to British sweeps. At the time, very young boys were sent down dark, soot-encrusted chimney flues to clean them. Because of the lack of appropriate protective clothing and limited personal hygiene, these boys were constantly exposed to soot, whose carcinogenic properties were definitively demonstrated more than a century later. Pott's findings contributed to the Chimney Sweeps Act of 1788, which stated that no

Figure 2.1 A chimney sweep often started working at a very young age. Constant exposure to soot and other carcinogens caused scrotal cancer.

boy under the age of eight years could serve as an apprentice to a chimney sweep. This represented the first link between epidemiology and public policy. It would be many years before fines and licensing made the lives of all chimney sweeps somewhat healthier. In 1875, the Chimney Sweepers Act was passed, imposing a licensing system that was enforced by the police. Scrotal cancer diminished among sweeps after the passage of the 1875 act.

James Lind (1716–1794) was a Scottish physician who in 1747 conducted the first clinical trial, identifying that eating citrus fruits was an effective remedy for scurvy among sailors at sea. Scurvy results from a deficiency in Vitamin C and causes general weakness, anemia, gum disease, and skin hemorrhage. Symptoms of advanced scurvy include poor wound healing, jaundice, neuropathy, and, in some cases, death. Scurvy was often found in sailors who were away from sources of fresh food, particularly fruits and vegetables, on long voyages.

Lind made his discovery by separating a larger group of sailors with scurvy into smaller groups, administering different treatments to each group. The treatments varied considerably and included vinegar; cider; sea water; a concoction of nutmeg, garlic, and mustard seed among other ingredients; and oranges and lemons. The group that received the citrus fruit treatment recovered the most quickly. Knowledge of how to treat scurvy is credited with helping the British defeat Napoleon and the French navy at Trafalgar. Lind's epidemiologic work led the British navy to require that limes or lime juice be included in the diet of seamen, resulting in the nickname of "limeys" for British sailors.

Smallpox, a contagious virus characterized by a rash, high fever, malaise, headaches, body aches, and vomiting, routinely devastated populations across history. Death and disfigurement rates from smallpox were high, making the disease a prime target for prevention efforts. As early as 590 BCE, the Chinese deliberately introduced weaker strains of smallpox in susceptible people in order to protect them against stronger strains of the disease. This process is called **variolation**.

Edward Jenner (1749–1823) was a British doctor who was intrigued by the popular notion in farming communities that people who caught cowpox from their cows did not catch smallpox. Cowpox is a mild viral infection found in cows. Blister-like pocks appear on the udders of the cows. The disease was transmissible to humans and often milkmaids developed a few of these lesions, usually on their hands if they were exposed to a cow with cowpox. In 1796, Jenner took cowpox material from the hand of a milkmaid and inserted it into the arm of a local farm boy, who had no known immunity to smallpox. He developed a mild case of cowpox and recovered completely. Several weeks later, to test the child's

Figure 2.2   Edward Jenner vaccinates his child against smallpox in this late 19th century engraving.

immunity to smallpox, Jenner inoculated him with material from a smallpox pustule. The child did not develop smallpox. After several months passed, Jenner again exposed the child to smallpox, and, once again, the boy did not develop the disease.

Jenner was a controversial figure because of his methods, but his contribution to stopping the spread of this deadly disease was significant. Jenner's discovery led to the development of the field of vaccinology, which takes its name from the Latin word for cow, *vacca*.

Smallpox is now **eradicated** after a successful worldwide vaccination program. The last case of smallpox in the United States was in 1949. The last naturally occurring case in the world was in Somalia in 1977. Routine vaccination against smallpox has ended because it is no longer needed for prevention.

Ignaz Semmelweis (1818–1865) is known as the father of infection control. In 1847, during his appointment as an obstetrics assistant in a hospital in Vienna, Austria, Semmelweis observed high levels of puerperal, or childbed, fever in mothers delivered by physicians and medical students, as compared to women delivered by midwives or midwife trainees. His investigation found that often medical students and physicians would attend births immediately after conducting autopsies. He hypothesized that students and doctors with putrefied cadaver material on their hands were conducting pelvic exams, thus exposing the mothers and babies to dangerous bacteria that caused childbed fever. Midwives did not conduct autopsies; thus, their patients were not exposed.

Acting on his hypothesis, Semmelweis instituted a mandatory hand-washing policy for medical students and doctors. The mortality rates for women delivered by students and doctors fell to the same rate as the midwives. When he began requiring the cleaning of instruments, the mortality rate dropped even further. Unfortunately, the prevailing theory of disease transmission at the time was the theory of **miasma**, which blamed the spread of disease on bad air, or miasmas. Thus, the new hospital ventilation was credited for the improvements in mortality rates, and the importance of hand washing was dismissed until decades after Semmelweis's death when Louis Pasteur and Robert Koch's work established evidence of germ-carrying microorganisms.

John Snow (1813–1858) is considered the father of modern epidemiology. He was an English anesthesiologist who developed several key epidemiological methods that remain valid and are still in use today. He is best known for his work on cholera, a waterborne illness. Over the centuries, cholera, caused by the *Vibrio cholerae* bacterium, inspired great fear because of the dramatic symptoms and mortality that it causes. Untreated cholera outbreaks can kill more than half of affected cases. The **fatality rate** is often less than one percent when the disease is treated. Snow believed that cholera was transmitted by contaminated water. This was in direct opposition to the current thinking of the time, which held that disease was transmitted by a miasma (or bad air or noxious gas) that came from rotting organic material. In Snow's time as in Semmelweis's, the connection between microorganisms and disease had not yet been ascertained.

During the 1848–1849 cholera epidemic, Snow, along with others from the London Epidemiological Society, advised the government on ways to fight the disease. Because of the symptoms of the disease, which attacks the gastrointestinal tract, Snow theorized that it was caused by something in the water and was likely spread by contact with feces and

Figure 2.3   John Snow, father of modern epidemiology.

Figure 2.4   Map drawn by Dr. John Snow of cholera cases clustering around the Broad Street pump during the London cholera epidemic of 1854.

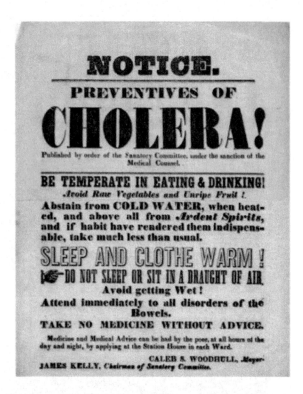

Figure 2.5 This poster suggests that miasma or bad air was the cause of cholera. The germ theory of disease developed after the invention of the microscope.

Figure 2.6 Louis Pasteur experimenting in his laboratory.

soiled clothing. This, of course, was directly at odds with miasma theory.

Snow is most famous for his work during another cholera outbreak in 1854 that centered around the Broad Street pump, a water pump located in the Soho neighborhood of London. He interviewed the residents of the area and graphed and mapped what he knew about the cases to pinpoint the pump as the possible source of the contamination. Further, he recommended a public health measure to prevent disease with the removal of the Broad Street pump handle.

Snow's "Grand Experiment" also took place during the epidemic of 1854. Two different water companies supplied water from the Thames River to houses in the same area of London. In 1852, the Lambeth Company relocated its sources of water to a less polluted portion of the river, while the Southwark and Vauxhall Company continued to draw water from a section of the Thames that was frequently polluted with sewage. Snow noted that during the cholera outbreak of 1854, those residents served by the Lambeth Company had fewer cases of cholera than residents served by Southwark and Vauxhall. Since the main difference in the individuals affected was where they obtained their water, Snow concluded that the water was the source of contamination.

Snow's work contributed to the development of many approaches still used today in epidemiology. Working in the field to interview subjects, mapping and graphing, and using data tables to describe infectious disease outbreaks are all important tools in the modern epidemiologist's arsenal of weapons against disease.

William Farr (1807–1883) was appointed "Compiler of Abstracts" in England in 1839. In this capacity, he developed a more sophisticated system for codifying medical conditions. He also examined possible linkages between mortality rates and population density and promoted the idea that some diseases,

Table 2.1. The Henle-Koch Postulates.

1. The organism must be observed in every case of the disease.
2. The organism must be isolated and grown in pure culture.
3. The pure culture must, when inoculated into a susceptible animal, reproduce the disease.
4. The organism must be observed in, and recovered from, the experimental animal.

especially chronic diseases, may have a **multifactorial etiology,** with many factors contributing to the development of disease. Farr also extended the use of vital statistics and developed a modern vital statistics system, much of which is still in use today. Interestingly, he was one of the most vocal adherents to miasma theory, and publicly challenged John Snow's theory that cholera was a germ and not transmitted through air. Years after Snow's death, Farr publicly acknowledged that the miasma theory was incorrect, based on his analysis of the cholera death rate data.

After the advent of the microscope, it was possible to conduct more formal experiments on the relationship between germs and diseases. Many of these experiments were conducted by **Louis Pasteur** (1822–1895). Popularly known as the father of microbiology, Pasteur is best known for the discovery of a process that is now called pasteurization, which involves killing disease-causing microbes—particularly those in milk and wine—by heating them.

Pasteur's contributions to the understanding of the causes and prevention of disease were substantial. He is credited with the discovery of the pathology of puerperal fever, rabies, and anthrax. He suggested sanitizing methods that greatly reduced deaths from puerperal fever and developed vaccines for rabies and anthrax.

Additional work on germ theory was conducted by **Jakob Henle** (1809–1885) and his student, **Robert Koch** (1843–1910) (see Table 2.1). Their work essentially suggested a mechanism by which an organism could be associated with a particular disease.

Koch used the microscope to see the microbes that cause tuberculosis and cholera. He used photography to take the first pictures of microbes in order to show the world that microorganisms do, in fact, exist, and that they are what causes diseases. Koch, following his postulates, showed that anthrax was transmissible and reproducible in mice. He also demonstrated that the anthrax bacillus was the only organism that caused anthrax in a susceptible animal. With Pasteur, Koch firmly established support for the germ theory of disease, which is, of course, the model still used today in infectious disease epidemiology.

MODERN EPIDEMIOLOGISTS

Janet Lane-Claypon (1877–1967) made significant contributions to the field of maternal and child health. In 1912, she published a **cohort study** showing that breast-fed babies gained more weight than those fed cow's milk. She also developed the first large **case-control**

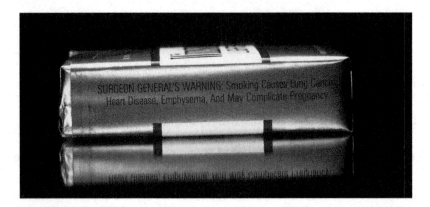

Figure 2.7   The work of Doll and Hill eventually led to this type of labeling on cigarette packages in the United States.

study (see Chapter 8) investigating whether women with a history of breast cancer differed from women with no history of breast cancer. Lane-Claypon conducted a groundbreaking study that showed that early diagnosis of breast cancer was related to increased survival. Her additional studies also demonstrated that breast cancer risk was greater for women who did not have children, who married at a later-than-average age, and who did not breast-feed.

After World War II, many countries experienced an increase in deaths attributable to lung cancer. Some of the first epidemiological studies assessing the association between cigarette smoking and lung cancer were conducted by **Richard Doll** (1912–2005) and **Austin Bradford Hill** (1897–1991) in the United Kingdom.

Figure 2.8   In addition to communicable diseases, many epidemiologists study the social determinants of health. This playground at a housing complex in Riga, Latvia illustrates how where one lives influences one's health. Children living here may not get much exercise because they do not have a safe environment in which to play. Consequently, they may be more likely to develop obesity and other related health disorders compared to children who have access to well maintained play areas.

In 1947, the team conducted a case control study, comparing smoking history differences between those with lung cancer and those without the disease. They found that the risk of lung cancer was directly related to the number of cigarettes smoked. A cohort study of British male physicians was conducted beginning in 1950, with questionnaires sent to determine their smoking habits. The study continued for 25 years, eventually demonstrating a link between smoking, heart disease, and an array of other serious diseases. Results from the earlier part of the cohort study, clearly establishing the link between smoking and lung cancer, led to the 1964 publication of the U.S. Surgeon General's report, *Smoking and Health*, identifying the deleterious effects of smoking upon health for the American public.

Michael Marmot (1945–) conducted the groundbreaking Whitehall II studies, which examined the relationship between socioeconomic status and health in government workers in Great Britain. This **longitudinal study**, which continues today, found an inverse relationship between the level of employment and health, such that those in jobs with lower salaries were more likely to have poorer health than those in better-paid positions. There were also differences in health risk behaviors such as smoking, diet, and exercise based on income level, with those making less money more inclined to engage in more risky behaviors. The discovery of this social gradient of health has led to understanding the need to study the social determinants of health—the conditions in which people are born, grow, live, work, and age—and to think about human health beyond infectious disease.

3 EPIDEMIOLOGICAL CONCEPTS

Epidemiology is the study of the distribution and determinants of health and disease, injuries, disability, and mortality in populations. The results of epidemiologic studies are applied to the prevention and control of health problems in populations. But what exactly does that mean? This chapter reviews the basic terms and concepts used by epidemiologists.

Distribution refers to the fact that the occurrence of diseases and other health outcomes varies in populations, with some subgroups of the populations more frequently affected than others. In epidemiology, we study how diseases are distributed across populations so that recommendations on prevention and control can be made based on evidence from the affected populations.

The epidemiological triangle (see Figure 3.1) shows the interaction and interdependence of agent, host, environment, and time as used in the investigation of diseases and epidemics. Agents are related to determinants in the definition of epidemiology. A determinant is any factor that brings about change in a health condition. Related to determinants are exposures, which pertain either to contact with a disease-causing factor or to the amount of the factor that impinges upon a group or individuals. Infectious agents are classified according to their size, structure, and physiology and include bacteria, viruses, rickettsia, helminthes, fungi and yeast, protozoans, and prions.

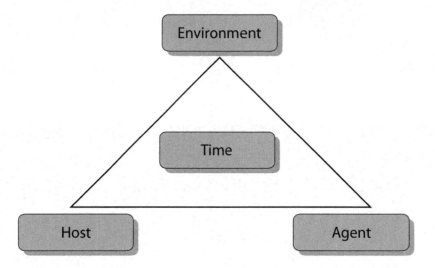

Figure 3.1 Epidemiological triangle.

Infections are caused by the entry and multiplication of microorganisms and parasites in the bodies of humans and animals.

A **host** is an organism, usually a human or an animal, who harbors a disease. Hosts can be **symptomatic cases** or people who have apparent signs of the infection. For example, an individual with a runny nose and a deep cough has obvious signs of influenza.

Carriers are people who harbor the infectious agent. They don't show any signs of infection, but are capable of transmitting the disease. The most famous example of a carrier is Mary Mallon, also known as Typhoid Mary. Mallon was a New York City cook. In 1906, six members of the household in which she was employed became ill with typhoid. She was identified as the first healthy carrier of the disease, shedding *Salmonella typhi* bacteria in her feces, but not showing any symptoms of the illness herself. She ultimately infected 22 people, including a little girl who died.

Mallon was quarantined when she refused to quit working as a cook, then released when she promised not to seek employment involving handling food. She was found to be working again as a cook after eluding authorities for five years, and was sent to enforced isolation on a remote island in the East River in New York City. Mallon died in 1938 after more than 26 years in quarantine.

Zoonoses are vertebrate animal carriers, often with complex life cycles and multiple hosts and insect vectors. Over the past several decades, more than 60% of newly identified infectious agents have come from pathogens originating from animals or animal products. A common example of zoonoses is rabies, which occurs when a rabid animal bites a human. Rabies is almost always fatal, but if a bitten human receives postexposure antirabies vaccinations (prophylaxis) prior to showing any signs of the disease, the odds of recovery are greatly improved.

Finally, disease can be transmitted from an inanimate object called a **fomite**. Certain microbes are free living in the environment in food, water, soil, air, and other inert substances. For example, tetanus can be found in soil, dust, and manure. The bacillus enters the body through breaks in the skin, generally through cuts and puncture wounds caused by a contaminated object.

Environment includes those surroundings and conditions external to the human or animal that cause or allow disease transmission. Within the environment is the **reservoir**, or the normal habitat (living or nonliving), in which an infectious agent lives, grows, and multiplies

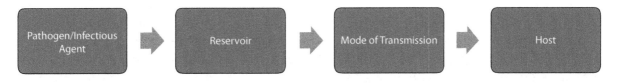

Figure 3.2   Chain of infection.

and is dependent on for its survival in nature. Without the proper environment, the agent cannot multiply and spread. Humans often serve as both reservoir and host.

There is a close association between the epidemiological triangle and the chain of infection, which relates to **time**. Time accounts for the incubation period (the time between the host first encountering the pathogen and when signs and symptoms first appear); life expectancy of the host or the pathogen; and duration of the course of the illness or condition. Time is often a key factor in trying to pinpoint the cause of an outbreak or epidemic through the use of epidemic curves. An **epidemic curve** involves graphing the distribution of cases by time of onset.

In conjunction with the epidemiology triangle, it is important to understand the chain of infection (see Figure 3.2) in order to arrest infection and minimize transmission. Public health officials often attack the reservoir of an agent at the environment leg of the epidemiological triangle. The chain of infection can be broken by destroying the pathogen in the environment in which it lives. For example, airplanes are often used to spray the watery breeding places (environment) of mosquitoes in order to kill the larvae thus preventing the transmission of mosquito borne diseases like malaria, yellow fever, and West Nile virus.

One of the goals of science of epidemiology is to **quantify** the relationship between a disease and exposure/risk factor in a specific population. The three most commonly used measures of association between a disease and risk factor are:

- Relative risk
- Odds ratio
- Prevalence ratio.

The **relative risk**, also known as the risk ratio, compares the risk of disease in the group with the exposure to the risk of disease in a group without the exposure. The relative risk can take on values that range from zero to infinity. If the relative risk equals 1.0, we say that the exposure has no impact on the risk of disease. If the relative risk is greater than 1.0, we conclude that the exposure is associated with more disease. On the other hand, if the relative risk is less than 1.0, we can conclude that the exposure is associated with a decreased occurrence of disease. In other words, if the relative risk is less than 1.0, the exposure protects against disease.

Similar to the relative risk, the **odds ratio** is a comparison of the occurrence of exposure among a group with the disease to those without disease. The interpretation of an odds ratio is similar to that of a relative risk. If an odds ratio equals 1.0, we conclude that there is no association between the disease and risk factor. If the odds ratio has a value greater than 1.0, we conclude that the odds of exposure are greater among the group with the disease compared to the group without the disease. An odds ratio less than 1.0 tells us that the exposure is more common in the group without the disease.

The third measure of association is a **prevalence ratio**, which compares the prevalence of disease among those with the exposure to the prevalence of disease among those without the exposure. If the prevalence ratio equals 1.0, we conclude that there is no association between disease and exposure. If the prevalence ratio is greater than 1.0, we conclude that the disease is associated with the exposure. On the other hand, if the prevalence ratio is less than 1.0 we conclude that the exposure is protective against disease.

Each of these measures of association will be discussed in greater detail when we discuss the different epidemiological study designs in the chapters to follow.

DISEASE TRANSMISSION

To propagate or spread, an infectious agent has to leave one host and go to another through one of the six portals in the human body: respiratory tract; conjunctiva; urogenital tract; gastrointestinal tract; skin; or placenta. **Transmission** is any mechanism by which an infectious agent is spread, and is essentially how an infectious agent bridges the gap between portals. Agents have a preferred portal: for example, the measles virus is inhaled into the respiratory system, whereas *E. coli* bacteria enter the body through the gastrointestinal tract.

Transmission can occur through **direct contact**, which is person-to-person physical contact such as touching with contaminated hands, skin-to-skin contact, kissing, direct droplet spread by sneezing or coughing, or sexual intercourse. Blood transfusions or infections that cross the placenta from mother to child are also forms of direct contact. **Indirect transmission** occurs when pathogens or agents are transferred or carried by some intermediate item, organism, means, or process to a susceptible host, resulting in disease. Indirect transmission may be vehicle borne, vector borne, or airborne.

Vehicle-borne transmission occurs through inanimate intermediates, including food and water, clothes, bedding, cooking utensils, and medical equipment. **Vector-borne transmission** occurs when the agent is carried by a live intermediary, such as an insect or an animal to the susceptible host. The agent may multiply in the host, or it may multiply in an intermediate host. **Airborne transmission** can occur through spores, dust particles, and through very small suspended droplets that enter the respiratory system.

Once we understand how a disease is spread, we can select the proper control method. Direct sources of transmission can be interrupted by blocking contact with the source. Indirect transmission is more challenging and may require interventions such as ventilation, sterilization of equipment, or proper food storage and sanitation.

DYNAMICS OF TRANSMISSION

Disease can spread in a variety of ways. **Common vehicle spread** is the spread of an agent through a common source—for example, through the air, water, food, or drugs. **Serial transfer** is transmission from human to human, human to animal to human, or human to environment to human in a sequence (measles, STDs, AIDS).

ENDEMIC, EPIDEMIC, AND PANDEMIC

When a disease or infectious agent is usually present in a community, a geographic area, or a population group, it is said to be **endemic**. We expect to see a certain amount of disease in a given situation. In 2013, malaria was endemic in 97 countries and territories; thus, we expect to see a particular number of malaria cases in any of these countries.

An epidemic is the occurrence of disease or other health-related events clearly in excess of what is normally expected. In malaria-endemic countries, this could be an increase in the number of cases of the disease in excess of what is normally seen in that region at that time of the year. An **epidemic** could also simply be one case of a disease. For example, in a country like the United States, where polio has been eradicated, it would be a public health concern if even one case of polio occurred.

An epidemic occurring worldwide or over a very wide area, crossing international boundaries, and usually affecting a large number of people is considered to be a **pandemic**. The 1918 Spanish influenza was a pandemic because more than one third of the world's population became ill, with about 50 million people dying from the disease. More people died from the Spanish flu than died during World War I.

Epidemics are commonly divided into **common source epidemics** and **propagated epidemics**. **Common source epidemics** occur when there is a pronounced clustering of cases of disease that occurs within a short time (i.e., within a few days or even hours) due to exposure of persons or animals to a common source of infection such as food or water.

Common source epidemics can be divided into point source epidemics and continuous source epidemics. **Point source epidemics** are defined by the exposed developing the disease very quickly, often over one incubation period. Food-borne illnesses are often point source epidemics, with everyone who consumes tainted food at an event becoming ill fairly rapidly. **Continuous source epidemics** occur when exposure to a source is prolonged over an extended period of time. A community continuing to drink from a contaminated water supply would result in a continuous source epidemic. Until access to the contaminated water is cut off, people will continue to become ill.

Propagated epidemics arise from infections being transmitted from one infected person to another. Transmission can be through direct or indirect routes. These host-to-host epidemics typically rise and fall more slowly than common source epidemics. Examples of propagated epidemics include measles, tuberculosis, whooping cough, and influenza.

Mixed epidemics can occur when a common source epidemic is followed by person-to-person contact, and the disease is spread as a propagated outbreak. Individuals may contract salmonella from improperly cooked meat at a restaurant (common source epidemic). If those who become ill do not maintain good hand hygiene, scrupulously washing their hands after every trip to the bathroom, they may spread the disease to others via oral-fecal transfer (propagated epidemic). Consequently, the disease can spread to many others who did not eat at the original venue.

SUSCEPTIBILITY AND CASE DEFINITIONS

In order for an exposure to cause a disease, an individual must be **susceptible** to the disease. This means an individual is vulnerable to a disease because he or she has no resistance or immunity to the disease. Everyone has some form of **innate immunity**, or inborn barriers, to disease and infection. These include physical barriers like intact skin, mucosal lining, cilia, and the cough-and-gag reflex. Humans also have various chemical barriers like acidity in the stomach, various enzymes, lipids, and interferons that create a hostile environment for agents seeking to invade.

Resistance is further acquired through either previous exposure to the pathogen or by immunization against the pathogen, which is usually through vaccination. This is called **active immunization**. **Passive immunization** occurs less frequently and occurs when antibodies from other sources are given as post-exposure prophylaxis for diseases like rabies and hepatitis. Maternal antibodies are also passively transmitted across the placenta to provide some protection to newborn infants.

Typically, we think of individuals as immune, but groups have immune statuses as well. **Herd immunity** is the proportion of individuals in the population who are resistant to a particular disease. When herd immunity occurs, the disease has no chance to take a hold because most people are immune. Everyone does not need to be resistant, but a high percentage of the population is required to be. How high a percent is based on the contagiousness of the disease. Moderate levels of herd immunity can slow infection without completely stopping a disease, but this can have unintended negative consequences. If there is no herd immunity, people may contract the disease as an adult, when the outcomes and side effects are more risky. For example, if a pregnant woman contracts rubella, her unborn baby may have a variety of congenital deformities that could have otherwise been prevented if she was vaccinated as a child.

A standard set of criteria, or **case definition**, assures that cases are consistently diagnosed, regardless of where or when they were identified and who diagnosed the case. A **case** is a person who has been diagnosed as having a disease, disorder, injury, or condition. The first disease case in the population is the **primary case**. The first disease case brought to the attention of the epidemiologist is the **index case**. The index case is not always the primary case. **Secondary cases** are those persons who become infected and ill after a disease has been introduced into a population and who become infected from contact with the primary case.

A **suspected case** is an individual who has all of the signs and symptoms of a disease or condition, but has not yet been diagnosed via laboratory or other definitive testing methods. A suspected case is upgraded to a confirmed case as more information (such as laboratory results) becomes available to the diagnosing physician. The diagnosis is upgraded when all case definition criteria are met. The case is then classified as a **confirmed case**.

WHY DO EPIDEMICS OCCUR?

Epidemics occur when susceptible people travel to an endemic area and are thus exposed to a new disease and also when a new disease is introduced to an area. The Native American population was decimated when smallpox was introduced to the Americas by European explorers. New or unusual social, behavioral, and cultural practices can expose a group that was previously unexposed to a new pathogen. Host susceptibility and response can also be modified by immunosuppression, malnutrition, or other diseases like AIDS.

WHY DO EPIDEMICS END?

An epidemic can be stopped when one of the elements of the epidemiological triangle is interfered with, altered, changed, or removed from existence, so that the disease no longer continues along its usual mode of transmission and routes of infection. This can happen if everybody who was susceptible got the disease and either recovered with full immunity or died as a result of the infection. Individuals might learn to stay away from the source of the infection—for example, not drinking from a contaminated stream, or the source of contamination is eliminated. The use of vaccinations and other preventive measures, such as bed nets

Table 3.1   Epidemiological Transition

LEADING CAUSES OF DEATH	
1900	2011
1. Pneumonia	1. Heart Disease
2. Tuberculosis	2. Cancer
3. Diarrhea	3. Chronic Lower Respiratory Disease

Source: Centers for Disease Control and Prevention

and insecticide treatments for mosquito control to reduce malaria transmission, can help stop epidemics. Finally, as some germs pass from one individual to another, they change or mutate and become less capable of producing disease.

COMMON EPIDEMIC INTERVENTIONS

Many actions can be taken to slow or end an epidemic. Public health professionals can control the source of the pathogen by removing the source, removing people from exposure, inactivating the pathogen by sterilization, and by treating and quarantining infected people. Transmission can be interrupted through vector control or personal hygiene, typically involving strong reminders to wash hands after using the bathroom, changing diapers, or coughing and sneezing and before and after food preparation. Public health officials can suggest that individuals modify their behaviors or use barriers to modify the host response to the exposure. An example of this would be an educational campaign on the importance of using condoms to prevent sexually transmitted diseases.

EPIDEMIOLOGICAL TRANSITIONS

An **epidemiological transition** is a shift in the patterns of disease and death from primarily acute, infectious disease to chronic, lifestyle-based diseases. Epidemiological transitions are related to **demographic transitions**, which occur as countries develop and grow economically. There is a shift from high birth and death rates to much lower birth and death rates. Epidemiological and demographic shifts are generally parallel and generally occur as a country moves from an agrarian society to a more industrialized society. Epidemiological transitions can be attributed to improved medical technology, improved standard of living, birth control, improved nutrition, sanitation and vector control, and improvements in lifestyle.

Improved medical technology provides vaccinations and better methods of detecting and treating illnesses. An improved standard of living means people have more money, more education, and increased health care access. The availability of birth control allows mothers to have fewer children, spaced further apart, resulting in decreases in both infant and maternal mortality rates. Improved nutrition improves overall health and increases a person's ability to resist infectious disease.

Infectious disease decreases during an epidemiological transition because of better sanitation and vector control. Individuals have access to clean water, ways to dispose of excrement, and flies, mosquitoes, and other vectors are controlled. Finally, there are general improvements in lifestyle. Individuals have time to engage in regular exercise and are able to take advantage of fresher air and green spaces.

CHRONIC DISEASE EPIDEMICS

Chronic diseases are the leading causes of death in developed nations and represent a considerable burden of disease in many developing nations as well. **Chronic diseases** are generally less severe, but are of continuous duration and must be managed appropriately. Some diseases start with an infectious agent, but then become chronic. For example, Lyme disease is caused by a bacterium, *Borrelia burgdorferi*, and is transmitted through the bite of an infected blacklegged tick. It is fairly easily treated if caught early. If left untreated, infection can spread to joints, the heart, and the nervous system, causing chronic, long-term difficulties for the infected person.

The latency period is relatively long for chronic diseases, and most have several interrelated causes. This **multifactorial etiology** suggests that a combination of components may be required before the chronic disease occurs. Often, both environmental and genetic factors are needed; sometimes, one or the other is enough to begin the chronic disease process.

Multifactorial etiology also makes disease prevention difficult, as prevention efforts must begin well before any signs or symptoms of disease occur. To account for the complexities involved, a more advanced model of the triangle of epidemiology is used that better reflects chronic disease issues (Figure 3.3). This model goes beyond what is considered for infectious diseases to include health behaviors, lifestyle factors, environmental causes, and other social determinants of health that influence the development of chronic diseases.

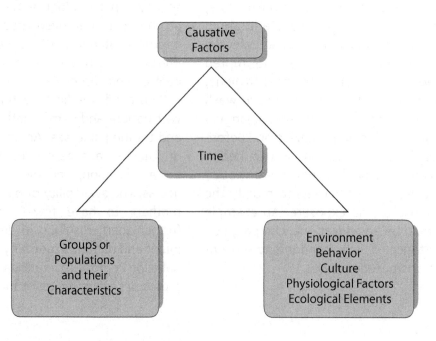

Figure 3.3   Chronic Disease Epidemiological Triangle

2X2 TABLES

One of the main goals of the science of epidemiology is to quantify the relationship between a disease and exposure/risk factor in a specific population. To do so, we start with a 2x2 table. A 2x2 table allows us to count the number of individuals with the disease and exposure of interest (Table 3.2).

The columns of a 2x2 table will include individuals with and without the disease of interest. The rows will count individuals based on their exposure status. The cells of our 2x2 table can be interpreted as follows:

a = the number of individuals who have the disease and are exposed

b = the number of individuals who do not have the disease, but have been exposed

c = the number of individuals who have the disease, but have not been exposed

d = the number of individuals who do not have the disease and have not been exposed

The sum of the cells a+b equals the total number of individuals who have been exposed; c+d equals the total number who have not been exposed. Similarly, the column total of a+c equals the total number of individuals with disease; b+d equals the total without disease.

The 2x2 table is the foundation of most epidemiological calculations. We will use the 2x2 table to calculate measures of association between disease and exposure. Understanding how to set up and interpret the data in a 2x2 table is imperative for all epidemiologists.

Table 3.2   The 2x2 Table for Epidemiological Studies

	Disease +	Disease −	
Exposure +	a	b	a+b
Exposure −	c	d	c+d
	a+c	b+d	N = a + b+ c + d

EXERCISES

1. A study was conducted to determine if exposure to pesticides on food is associated with mouth cancer. In our study we found that of the 200 individuals who were exposed to pesticides, 150 have mouth cancer. Of the 300 individuals who were not exposed to pesticides, 150 have mouth cancer.

    a.   What is the exposure? What is the disease?

    b.   Complete the corresponding 2x2 table.

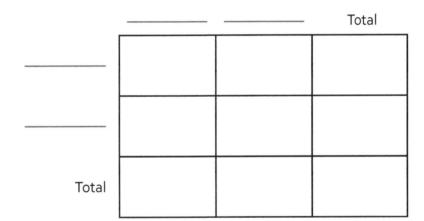

    c.   Explain the value of each cell.

    d.   How many individuals have mouth cancer?

    e.   How many individuals are there without mouth cancer?

2. A study was conducted to determine if smoking cigars is associated with tongue cancer. Of the 500 individuals with tongue cancer, 450 were cigar smokers; of the 1,000 without tongue cancer, 450 were cigar smokers.

    a.   What is the exposure? What is the disease?

    b.   Complete the corresponding 2x2 table.

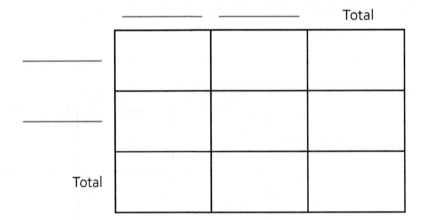

    c.   Explain the value of each cell.

    d.   How many individuals were cigar smokers?

    e.   How many individuals were not cigar smokers?

4 MEASURES OF DISEASE

Morbidity is the measurement of disease in a population. Morbidity can be measured using a rate or proportions (see the Mathematical Concepts chapter for a refresher on these concepts). **Rates** indicate how fast the disease is occurring in a population, whereas **proportions** tell us what fraction of the population is affected by the disease. The measurement of disease frequency in a population starts very simply with case counts. We can count the proportion of the population with a disease risk factor or a disease at a particular point in time, which is **prevalence**. This tells us the extent of the disease within the population of interest. Prevalence data is often gathered at the starting point of a study, also called the baseline time point. Prevalence is:

$$\frac{\text{The number of EXISTING cases of a disease at a specified time}}{\text{Total population at the specified time}}$$

Point prevalence is the prevalence of a disease in a population at a single *point* in time.

Period prevalence is the prevalence of a disease in a population over a specified *period* of time (thus it includes cases at the start of the period and any subsequent new cases), therefore:

Period prevalence = point prevalence + incidence

For rare diseases with short duration, period prevalence is approximately equal to the incidence.

For example, we might be interested in the prevalence of hysterectomy in a population of women ages 50 to 65. We would recruit 1200 women and determine that 97 had hysterectomies. We can calculate the prevalence:

$$\text{Prevalence} = \frac{\text{no. of existing cases}}{\text{no. of people in population}}$$

$$= \frac{97 \text{ women}}{1200 \text{ women}} = .08 = 8\%$$

There are several factors that increase prevalence. Prevalence is increased by a longer duration of a disease. For example, the prevalence of tuberculosis is increased when more individuals are diagnosed with drug-resistant forms of the disease, which take much longer to cure. The lives of patients with a disease may also be prolonged with a treatment for the disease, serving to increase the prevalence. Individuals diagnosed with HIV are now able to live longer, healthier lives than in previous decades due to the development of a variety of antiretroviral drugs. These individuals still have HIV, and thus are existing cases, and so are counted in the prevalence of the disease. An increase in new cases (greater incidence) can also increase prevalence. This is clearly illustrated during influenza season, when a large number of people become ill at roughly the same time.

Prevalence also increases when a number of individuals with existing cases of a disease move in large numbers into a particular region. An out-migration of healthy people will also cause prevalence to increase. We might see either of these scenarios in relation to demographic shifts in populations, which are often related to economic factors. For example, as the concept of active adult retirement communities becomes more popular, older individuals (age 55+) may be attracted to a particular area. Age is a strong predictor of disease, and thus prevalence of diabetes or heart disease may increase in these areas with active adult retirement communities. Alternatively, an out-migration of healthy people may occur as an economic downturn affects an area and those healthy enough to relocate move in search of employment.

Finally, as medical technology improves, an increase in prevalence may be seen. This is a result of diagnostic and screening tests becoming more refined and better able to pinpoint diseases at earlier stages, contributing to better and more thorough reporting of the prevalence of the disease in question.

Prevalence decreases when the disease is of shorter duration Either people recover from the disease and are immune, or they reenter the vulnerable population. Prevalence will also decrease when there is a higher **case-fatality rate** and individuals with the disease die from it. If a vaccine is developed, natural immunity occurs, or transmission is curtailed, there will be a reduction in new cases or decreased incidence, and prevalence will decline. Improving cure rates of the disease also reduces prevalence, as does an in-migration of healthy people.

INCIDENCE

Incidence is the number of new cases of a disease that occur during a specified time in a population at risk for the disease. Incidence data can be reported in a number of ways, including as decimals, percents, and using what is called a unit multiplier. The formula is:

Number of NEW cases of a disease occurring in a population during a specified period of time

---

Number of persons at risk of developing the disease during that period of time

As can easily be seen, the formula for the calculation of incidence is remarkably similar to the calculation of prevalence. The key difference is that the numerator of the incidence calculation involves only new cases of disease. This means that individuals who currently have the disease or condition of interest are removed from the calculation. Incidence is a measure of events, marking the move from disease-free to diseased status. Thus, it is considered a measure of risk; prevalence is not a measure of risk.

The denominator of incidence reflects the number of individuals who are at risk for developing the disease. This means that any individual who is included in the denominator must have the potential to be a part of the group that is included in the denominator. Consequently, if we are looking at the incidence of prostate cancer in a particular population, only men can be included in the denominator because women are not at risk for developing prostate cancer.

For example, suppose we wanted to determine the incidence of uterine cancer in women ages 50 to 75. We recruit a cohort of 1200 women. Of these women, 97 have had a hysterectomy, which includes the removal of the uterus. Without a uterus, these women are not at risk of developing uterine cancer; thus, our denominator is reduced by these 97 women who are not at risk. Our denominator is calculated based on: 1200 recruited − 97 no longer at risk = 1103. These 1103 at-risk women are followed over the next year, and 27 onsets of uterine cancer are observed.

$$\text{Incidence} = \frac{\text{no. of new cases}}{\text{no. @ risk}} = \frac{27 \text{ women}}{1103 \text{ women}} = .02 = 2\%$$

The incidence in the population is .02, or 2%.

Another issue to consider with denominators in incidence calculations is time. A time period must be expressed in order for incidence to be a measure of risk. The actual time period is inconsequential—we can follow a group for a week, a month, a year, or ten years and still calculate incidence. What is important is that all the individuals identified as at risk at the beginning of the time period are followed for the specified period of time. This results in **cumulative incidence**.

In food-borne outbreak investigations, the **attack rate** is synonymous with incidence. Symptoms of a food-borne illness generally occur very quickly, within hours to days after eating a contaminated food. For example, the incubation period for *E. coli*, a common food-borne pathogen, is one to ten days. The symptoms include severe diarrhea, vomiting, and abdominal pain. It is doubtful that a person has harbored the infection for several months; thus, it is very likely a new case of the disease and therefore represents incidence. Cases that develop months later are not considered part of the initial outbreak.

ISSUES WITH MORBIDITY DATA

There are some problems with incidence and prevalence measures. Problems with numerators, which are based on those with the disease, include issues with how a disease is defined and where cases are obtained. If differing case definitions are used across studies, it makes it very difficult to make comparisons or ascertain the actual impact of a particular disease. It can also be difficult to find individuals with the disease. Do we use existing data, or do we conduct a new study? Are we sure of the accuracy of existing data? If we do a new study, do our cases really know for certain they have the disease, or, for that matter, that they are truly free of the disease and actually belong in the denominator?

Problems with denominators include issues with how particular populations are counted. If we are looking at ethnicity, how is that data captured? Further, everyone who is at risk in the population of interest must be included in an incidence calculation. How can we be sure that we've gotten everyone? Can we realistically include everyone?

EXERCISES

Please circle the correct answer:

1. Incident cases or prevalent cases: The Organic Oasis Food Company keeps careful records of on-the-job injuries. Every worker who sustains an injury in the factory must report the injury to the Occupational Safety Office immediately. There are 500 workers at the factory, and there were 5 injury reports filed the third week of October.

2. Incidence, period prevalence, or point prevalence: The health department in Moorea is concerned about low-weight-for-height children under three years of age. There are 200 children under three years of age in the village. When the health department measured all 200 children on June 12, they found 19 children had scores that indicated low weight for height.

3. Incidence or prevalence: An outbreak of vomititus occurs every spring in North Adams County. Within 12 to 24 hours of infection with vomititus, everyone who has been infected develops characteristic, easily diagnosable symptoms that resolve after 3 to 5 days. North Adams County has 2000 residents. The first 3 vomititus cases of the year were diagnosed the week of March 28.

4. Incidence or prevalence: Residents at the Jonquil Retirement Apartments were asked: "Have you ever been told by a doctor that you have high blood pressure?" Of the 160 residents, 27 answered "yes."

5. Incidence, period prevalence, or point prevalence: Bowen's disease is a common childhood infection. The duration of symptoms varies from one day to several weeks. No immunity to Bowen's disease is conferred by infection, so people who recover from the illness are immediately susceptible to another bout of the infection. On December 1, 35 of the 1000 students who attend Green Springs Elementary School were home sick with Bowen's disease. During the rest of December, 145 students missed at least one day of school due to this disease. The total number of students who missed at least one day of school due to Bowen's disease was 180, or 18% of the student body. Is the 18% the incidence, period prevalence, or point prevalence?

In a study concerned with the possible effects of air pollution on the development of chronic bronchitis, the following data were obtained:

> A population of 9000 men, age 45 years, was examined in January 1990. Of these men, 6000 were exposed to high levels of air pollution, and 3000 were exposed to low levels of air pollution. At the physical examination, 90 cases of chronic bronchitis were identified, including 60 cases among the high-exposure group.
>
> All of the men who were initially examined and did not have chronic bronchitis were followed by repeat examination over the next five years. These follow-up exams identified 268 new cases of bronchitis in the total group, including 30 cases among the low-exposure group.

6.   The prevalence of chronic bronchitis in January 1990 were:

	NUMERATOR	DENOMINATOR	PREVALENCE
High-Exposure Group			
Low-Exposure Group			
Total Group			

7.   Is this a measure of point prevalence or a period prevalence?

8.   The incidence of chronic bronchitis over the five years were:

	NUMERATOR	DENOMINATOR	INCIDENCE
High-Exposure Group			
Low-Exposure Group			
Total Group			

According to the Centers for Disease Control and Prevention, Lyme disease is caused by the bacterium *Borrelia burgdorferi* and is transmitted to humans through the bite of infected blacklegged ticks. Typical symptoms include fever, headache, fatigue, and a characteristic skin rash called erythema migrans (bull's-eye rash). If left untreated, infection can spread to joints, the heart, and the nervous system. Repellents that contain 20% or more DEET (N, N-diethyl-m-toluamide) are recommended to prevent the disease.

A population of 6,175 northern Virginia residents who spent more than 15 hours per week working outdoors in wooded areas was recruited for a study to determine the effectiveness of DEET in preventing Lyme disease. Of these individuals, 3,125 used DEET sprays as recommended for Lyme disease prevention. The remaining individuals did not use DEET. During the physical examination, 191 cases of Lyme disease were identified, including 124 cases among the no-DEET-usage group.

All of the individuals who were initially examined and did not have signs or symptoms of Lyme disease were followed by repeat examination over the next year. These follow-up exams identified 259 new cases of Lyme disease in the total group, including 88 cases in the DEET-usage group.

9. Calculate the prevalence of Lyme disease in each study group at the beginning of the study.

	NUMERATOR	DENOMINATOR	PREVALENCE
DEET-usage			
No-DEET-usage			
Total			

10. Calculate the incidence of Lyme disease in each study group at the end of the follow-up year.

	NUMERATOR	DENOMINATOR	INCIDENCE
DEET-usage			
No-DEET-usage			
Total			

In a study concerned with alcohol use and liver cirrhosis, the following data were collected:

A population of 5000 individuals was examined in January 2010. Of this group, 2500 were considered heavy-drinkers (5 or more alcoholic drinks per day) and 2500 were moderate-drinkers (less than 5 alcoholic drinks per day). At the physical examination, 100 cases of liver cirrhosis were identified, including 65 in the heavy-drinker group.

All of the individuals who were initially examined and did not have liver cirrhosis were followed by repeat examination over the next two years. These follow-up exams identified 300 new cases of liver cirrhosis in the total group, including 95 cases among the moderate-drinker group.

11. Calculate the prevalence of liver cirrhosis in January 2010 were:

	NUMERATOR	DENOMINATOR	PREVALENCE
Heavy-drinker group			
Moderate-drinker group			
Total			

12. The incidence of liver cirrhosis over the two years were:

	NUMERATOR	DENOMINATOR	INCIDENCE
Heavy-drinker group			
Moderate-drinker group			
Total			

5 MORTALITY AND AGE ADJUSTMENT

Everybody dies, eventually. Obviously, there is great interest in how and why we die and what we can do to live a longer, healthier life. While the discussion of mortality statistics may seem to some to be, quite frankly, morbid, it is important to understand the rates at which people are dying and from what causes. Studying **mortality** (death) data quantitatively can pinpoint differences in the risk of dying from a disease for people in different categories and can help us determine **life expectancy**, or the number of years a person is expected to live after a given age. It allows us to answer questions related to prognosis that patients often ask. Mortality rates also tell us about disease severity if many people die from a particular cause. This helps decide priorities for clinical research and public health programs. Mortality rates can help determine if a treatment or a screening process is effective, as we would expect to see cases decline if the intervention is working. We can compare different treatments as well to discover which is more effective. Mortality data can also serve as a surrogate for incidence data if a disease is particularly virulent (highly pathogenic) and kills quickly. There are five main measures of mortality data—crude mortality rate, cause-specific mortality rate, case fatality rate (CFR), proportionate mortality rate (PMR), and years of life lost (YLL), also referred to as years of potential life lost (YPLL).

CRUDE MORTALITY RATE

The crude mortality rate is the number of deaths from all causes in a population divided by the number of persons in the total population. The formula for the crude mortality rate is:

$$\frac{Deaths\ from\ all\ causes}{Total\ population}$$

The unit is in per number of people and thus can be described in per 1000, per 10,000, or any unit that makes the most sense to the individuals who will be using the information.

Example: Of the 22,549 people living in the city of Maple Shade, 325 died of various causes during the year.

$$\frac{325\ deaths\ from\ all\ causes}{22,549\ population\ of\ Maple\ Shade}$$

$$=\ 0.014\ x\ 10,000$$

$$=140\ deaths\ per\ 10,000\ people$$

We simply multiply the .014 by 10,000 to provide a number that would be interpretable, giving us a crude mortality rate of 140/10,000 people in Maple Shade.

ADJUSTED MORTALITY RATES

Often we use mortality rates to make comparisons across geographic areas and between multiple populations (for example, between a group with a specific exposure and one without). Since the crude mortality rates are calculated by simply dividing the number of people who died by the total population, we are unable to compare rates when the underlying structures of the populations are different. For example, the age structure of the United States is different from Kenya or Nigeria. The United States has a longer life expectancy, as well as a slower rate of growth than both African countries. Nigeria and Kenya are both experiencing rapid growth, resulting in a large proportion of each country's population being less than 25 years old. The rates of death between the United States and Africa cannot, therefore, be compared using the crude death rate because of the differing population structures.

To deal with such a situation, most often we adjust the mortality rates to remove the effect of age so the populations are similar in structure.

Adjusting mortality rates for age allows us to answer the question:

If the age compositions of the populations of interest were the same, would there be any difference in mortality rates?

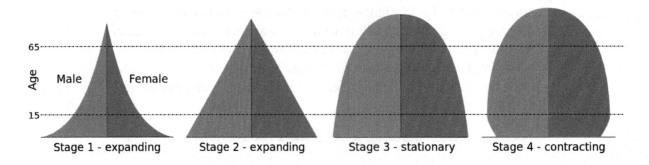

Figure 5.1 Illustration of differing population age structures.

There are two methods for adjusting mortality rates:

- direct standardization;
- indirect standardization.

Age Group	1900–1929 Death Rate (per 100,000)	1989–2009 Death Rate (per 100,000)
30–49	20	10
50–69	300	250
70+	450	400

DIRECT STANDARDIZATION

The direct standardization method for adjustment requires that we select a standard population first. This standard population could be *external*, the total population of a larger area. For example, census data could be used. Alternatively, an *internal* standard population, the sum of the two populations being compared, can be used.

Once a standard population (also called the reference population) has been established, we apply the age-specific mortality rates in the populations of interest to the standard population. This application of age-specific mortality rates to the standard population results in the count of expected number of deaths.

For instance, suppose we want to compare death rates during two time periods, 1900–1929 and 1980–2009, in a medium-sized town in rural Pennsylvania. The crude mortality rates (calculated as the number of people who died in the whole population) are:

165 per 100,000 in 1900–1929
175 per 100,000 in 1989–2009.

We can assume that the average life expectancy was lower during the period of 1900–1929. Additionally, we might assume that the underlying age structure during these two time periods is different.

The table below shows the age-specific rates for each time period.

By applying the rates of a standard population, we can calculate the expected number of deaths and age-adjusted mortality rate for each time period.

Age Group	Standard Population (a)	1900–1929 Death Rate (b)	1900–1929 Expected Deaths [(a*b)/100,000]	1980–2009 Death Rate (c)	1980–2009 Expected deaths [(a*c)/100,000]
30–49	500,000	20	100	10	50
50–69	300,000	300	900	250	750
70+	200,000	450	900	400	800
TOTAL	1,000,000		1,900		1,600

Direct Age-Adjusted Rates:
1900–1929: 1,900/1,000,000 = 0.00190 = 190 per 100,000
1980–2009: 1,600/1,000,000 = 0.00160 = 160 per 100,000

INDIRECT STANDARDIZATION

The indirect standardization process for adjustment is used when the age-specific death rates are unavailable. Using the indirect method results in the calculation of the standard mortality ratio (SMR).

$$SMR = \frac{\# \text{ observed deaths per year}}{\# \text{ expected deaths per year}}$$

The number of observed deaths is what is known. The expected number of deaths is calculated by applying the age-specific mortality rate in a standard population to the number of subjects in each age group. The age-specific expected number of deaths are then added together.

For example:

AGE	# WORKERS	POPULATION DEATH RATE	EXPECTED # OF DEATHS
20–29	Values known from study	Information from census or published literature	= # workers * death rate in population per age group
30–39			
40–49			
50–59			
60+			

AGE	# WORKERS	POPULATION DEATH RATE	EXPECTED # OF DEATHS
20–29	6000	0.0005	3
30–39	2000	0.0005	1
40–49	1000	0.002	2
50–59	600	0.02	12
60+	400	0.03	12
TOTAL	10,000		30

If the observed number of deaths is 50 per 10,000, our SMR would equal:

$$SMR = \frac{50}{30} = 1.67$$

If the SMR = 1, then the number of deaths observed were the same as expected from the age-specific rates in the standard population. If the SMR > 1, this indicates that more deaths were observed than expected from the age-specific rates in the standard population. If the SMR < 1, then we report that there were fewer deaths observed than expected from the age-specific rates in the standard population.

Both the direct and indirect standardization processes can be used to adjust for other variables, such as gender, smoking, and socio-economic status.

CAUSE-SPECIFIC MORTALITY RATE

The cause-specific mortality rate allows us to evaluate death rates related to a particular disease or condition. It is calculated by looking at the number of deaths from a particular cause divided by the number of persons in a total population. Cause-specific mortality rates can be narrowed down to look at particular subgroups of the larger population, as long as the denominator reflects that subgroup as well. For example, we can look at the number of deaths in women ages 30 to 45 from pancreatic cancer in a particular area as long as the denominator is the total number of women ages 30 to 45 living in our area of interest. Specific groups can be compared using a relative risk; thus, the death rate among women ages 30 to 45 from pancreatic cancer can be compared to the death rate among men ages 30 to 45 from pancreatic cancer. This number is typically also expressed in per number of people like the crude mortality rate, but it can also be expressed as a percent. The formula for the cause-specific mortality rate is:

$$\frac{number\ of\ people\ who\ died\ from\ the\ cause\ of\ interest}{total\ population\ during\ the\ time\ period}$$

Using our previous crude mortality rate example, of the 325 people who died during the previous year in Maple Shade, 12 died from myocardial infarctions (MI or heart attacks). Our population remains at 22,549. The cause-specific mortality rate would be expressed as

$$\frac{12\ died\ from\ MI}{22,549\ population\ of\ Maple\ Shade}$$

$$= .0005 \times 10,000$$

$$= 5\ per\ 10,000\ people$$

We multiply the .0005 by 10,000 to provide a number that would be interpretable, giving us a cause-specific mortality rate for myocardial infarctions of 5/10,000 people in Maple Shade.

We can calculate a cause-specific mortality rate for a subgroup. There are 1,645 women ages 30 to 45 in the city of Eastin. During the past year, 5 women (ages 30 to 45) died from pancreatic cancer in Eastin. This would be expressed as:

$$\frac{5 \text{ women ages 30 to 45}}{1645 \text{ women in Eastin ages 30 to 45}}$$
$$\text{who died from pancreatic cancer}$$

$$= .003 \times 1,000$$

$$= 3 \text{ per 1,000 women}$$

Since .003 is such a small number and likely has no meaning for most people, we can multiply it by 1,000, which gives us a pancreatic cancer cause-specific mortality rate of 3/1,000 in women ages 30 to 45 in Eastin.

CASE FATALITY RATE

The case fatality rate (CFR) is the number of deaths from a specific disease divided by the number of individuals in the population with the specific disease. This tells the measure of severity of a disease and is often expressed as a percent.

$$\frac{\text{Number of deaths from a specific cause}}{\text{Number of individuals in the population}}$$
$$\text{with the specific disease}$$

Example: If 375 children contracted chicken pox and 25 died from the disease, our CFR calculation would look like this:

$$\frac{25 \text{ died from chicken pox}}{375 \text{ contracted chicken pox}} = 6.66\%$$

Typically we express the case fatality rate as a percent because it is easier to understand.

CAUSE-SPECIFIC MORTALITY RATE VERSUS CASE FATALITY RATE

The cause-specific mortality rate and the case fatality rate are often confused because of their similar sounding names and the fact that they focus on a particular cause of death. An example to illustrate the difference:

Assume that in a population of 10,000 people, 20 people are sick with pancreatic cancer. In one year, 18 of the 20 people die from pancreatic cancer. The cause-specific mortality rate as a result of pancreatic cancer is 18/10,000 = .0018 or .18%. The case fatality rate as a result of pancreatic cancer is 18/20 = 0.9 or 90%. Thus, one could say that pancreatic cancer is not a leading cause of death in our population, but individuals who have pancreatic cancer are likely to die from it.

PROPORTIONATE MORTALITY RATE

The proportionate mortality rate (PMR) tells us what proportion of the dead died from a particular cause. It is calculated by taking the number of deaths from a particular cause divided by the total number of deaths in the population.

$$\frac{\text{Number of deaths from a particular cause}}{\text{Total number of DEATHS in the population}}$$

To help clarify, this is the only mortality rate calculation where everyone in the calculation is dead. This number is most often expressed as a percent. Thus, in our previous Maple Shade example, of the 325 residents who died in 2012, 75 died from stroke. Our PMR for stroke would be:

$$\frac{75 \; deaths \; from \; stroke}{325 \; deaths \; in \; maple \; shade} = .23 \; or \; 23\%$$

YEARS OF POTENTIAL LIFE LOST/YEARS OF LIFE LOST

Years of potential life lost (YPLL) or years of life lost (YLL) reflects the concept that death occurring at a younger age results in greater loss of future productivity than would death at a later age. This means that the younger the age at which death occurs, the more potential years of life are lost. Dying at a younger age is referred to as "premature mortality." This calculation is often used to highlight specific causes of death affecting younger age groups.

YPLL/YLLs can be calculated using individual level data or group level data. Each individual's YPLL is calculated by subtracting the person's age at death from the reference age, which is the standard life expectancy for the general population of interest. Only those who have died before reaching the standard life expectancy are included in the calculation.

Example: Assume the average life expectancy in Kenya is 57. If an infant dies at age 1, $57 - 1 = 56$ years of life lost. If an adult dies at age 50, $57 - 50 = 7$ years of life lost.

To calculate the YPLL for a particular population at a particular point in time, the individual YPLLs for all who died that year in that population are summed. All cause mortality and cause-specific mortality can be studied in this manner.

EXPLANATIONS FOR MORTALITY TREND CHANGES

Trends in mortality data may change for a variety of reasons, both statistical and real. From a statistical perspective, there may be errors in classifying the cause of death—for instance, someone may have lung cancer, but die of a complication related to lung cancer. Would that be considered a lung cancer death, as the individual might not have died of the complication if he or she did not have a primary diagnosis of lung cancer? Errors in determining the age at death are common, mostly in developing countries, where birth certificates may not be regularly issued. The cause of death may be reported incorrectly, so that the individual death is classified in the wrong category. These are all issues that would affect the numerator of a mortality calculation.

Statistical issues relating to the errors in counting a population influence the denominator of mortality calculations. Think what an undertaking it is to count the population of an entire country. In the United States, this is only done every ten years; consequently, the calculations toward the end of that time period are less reliable than those done at the beginning. There may also be issues with classifying people into demographic categories, particularly relating to race and ethnicity. Many people are multiracial, so in which category do they belong? They are not likely one category, and so much information is lost if individuals are not correctly classified.

Over and above the issues posed by statistical errors, there are real reasons for changes in mortality trends. There can be changes in incidence, with an increase or decrease in new deaths from a particular cause. If there are more new cases resulting in deaths, the mortality rates will go up. If there are fewer new cases of a disease and thus fewer people dying from the disease—which is more in line with what we wish to see in public health—then we would see a decrease in mortality rates.

We can also see changes in survivorship without a change in incidence, which would lead to a decrease in death rates. As treatment for certain diseases improves, fewer people die from the disease. In this scenario, the number of individuals diagnosed with the disease does not increase, but treatment is better, so survival rates increase and mortality rates decrease.

Changes in the age composition of a population may affect mortality rates. The greatest predictor of death is age; as a result, communities with a greater older population will have higher mortality rates. For example, as previously discussed, if an area builds a number of active adult (65+) retirement communities, attracting additional older adults to the community, the age structure of the population will change, and the crude mortality rate may go up. We may see the opposite occur if a community builds a new school or other places that attract families with young children. Recall from our previous discussion of age adjustment that the crude mortality rate may drop because of the influence of a great number of younger people in the population. Of course, mortality rates may change due to a combination of any of these factors.

EXERCISES

Takini is a community of 126,429 people (63,214 males). During 1997, there were 1000 deaths from all causes. A screening program for tuberculosis (TB) detected all 300 cases of TB in the town (200 males). During 1997, there were 60 deaths from TB (50 males). Assume that in Takini, all TB cases that do not result in death are cured shortly after treatment (that is, the annual screening program detects incident cases).

1. What is the crude mortality rate in Takini?

2. What is the cause-specific mortality rate due to tuberculosis?

3. What is the proportionate mortality rate (PMR) due to tuberculosis?

4. What is the overall case fatality rate from TB?

5. What is the case fatality rate from TB for women?

6. What is the case fatality rate from TB for men?

7. What is the case fatality rate from TB comparing women to men?
    (Hint: Men are the reference group, so they go in the denominator of the ratio).

8. What is the case fatality rate from TB comparing men to women?
    (Hint: Now women are the reference group so they go in the denominator.)

6

DESCRIPTIVE STUDIES

A **study design** is the blueprint that allows for an assessment of events and for statistical inference concerning relationships between exposure and disease and defines the domain for generalizing the results. Study designs can be divided up into descriptive studies and analytic studies. **Descriptive studies** observe the frequency of health-related states and provide a means of organizing, summarizing, and quantifying epidemiological data by person, place, and time. Descriptive studies do not test hypotheses, but they do provide useful information regarding the extent of a public health problem; thus, they are usually the first step in an epidemiological investigation. Descriptive studies are followed by analytic studies that generate hypotheses, examine associations, and attempt to find causal relations between exposures and outcomes.

Describing data by person tells us the frequency of a disease in a population and identifies who is most at risk. We can categorize by a number of variables, including demographic characteristics (gender, race, ethnicity, marital status, family status, and education), activities (occupation, activity level), and social conditions (housing, local community environment, access to health care). We can also look at the effects of beliefs, attitudes, culture, and the social environment.

Describing by place, for example, by city, state, country, birthplace, and place of employment provides geographic information regarding the extent of a health issue. This provides clues as to where an agent resides or multiplies or where a risk factor may be greater. Transmission and spread of disease can also be tracked geographically.

Describing by time gives us an idea of when a disease may occur. Vector-borne diseases like malaria tend to increase during the rainy season, when standing water increases the breeding area for mosquitoes. Other diseases, like influenza, increase during the colder months, when people spend more time together in enclosed environments. Knowing the interaction between person, place, and time helps us better understand the course of a disease and how best to combat it.

Epidemiological studies are typically divided into descriptive studies and analytic studies. Descriptive studies provide information about a disease or a condition. Using descriptive epidemiology, we can identify the extent of a public health problem and can communicate about disease in terms the public can understand. These types of studies also help provide clues to possible new diseases or adverse health effects. Descriptive studies identify the populations at greatest risk. The knowledge gained from descriptive studies assists in programming planning and resource allocation and identifies areas for future research.

There are four basic types of descriptive studies: case reports, case series, ecological studies, and cross-sectional studies.

Table 6.1   How We Use Descriptive Epidemiological Studies.

DESCRIPTIVE EPIDEMIOLOGICAL STUDIES ARE USED TO …
1. Identify extent of a public health problem
2. Identify populations at greatest risk of disease
3. Identify changes in disease or risk factors across time or space
4. Provide data for public health program planning
5. Guide public health resource allocations
6. Identify areas for future research

has the same outcome. They do provide evidence for larger scale studies because they allow for hypothesis generation. Case reports and case series are not true epidemiological studies, but they can lead to epidemiological studies to test hypotheses suggested by the case reports. A single case study does not prove cause, but may stimulate others to report similar cases.

CASE REPORTS AND CASE SERIES

A **case report** involves a profile of a single individual, whereas a **case series** includes a series or group of case reports from individuals with the same disease. There are no comparison groups in these types of studies, as everyone

ECOLOGICAL STUDIES

Ecological studies use data aggregated at a population level, using groups of people, not individuals. These studies measure ecological effects experienced by large groups of people. Because ecological studies consider group characteristics, average measures are used. Each data point in an ecological study represents the mean value for a population. Generally, a scatter plot is created and a correlation is calculated. Since ecological studies allow for comparisons among groups of people, different populations can be compared at the same time, or the same population can be compared at two different time periods.

Ecological studies do not allow for causal inference because the data is collected at the group level. The link between an individual exposure and an individual outcome cannot be clearly made. Associations found at the group level do not necessarily remain when the association is considered at the individual level. An **ecological fallacy** is assuming an exposure is causal because it is prevalent in the same population that has a high prevalence of a certain outcome.

Ecological studies are low cost, simple, and convenient. Caution should be used when interpreting the results of ecological studies because the data is typically collected for reasons other than assessing epidemiological associations. Nonetheless, they can provide important information to begin to assess etiological relationships.

In Figure 6.1, each red diamond represents the average air quality index (AQI) measurement and rate of chronic obstructive pulmonary disease (COPD) in several different populations. Assume that higher AQI indicates poorer air quality. We see a trend that suggests that as the AQI score increases, the rate of COPD increases. We can conclude that poor air quality is associated with an increased rate of COPD at the population level. This evidence does not prove that poor air quality causes COPD at the individual level.

CROSS-SECTIONAL/ PREVALENCE STUDIES

Cross-sectional studies, often called **prevalence studies**, are conducted over a short period of time. Data is collected at the individual level. There is no follow-up period. In this type of study, the exposure and outcome are measured at the same time. Cross-sectional studies are often conducted at the start of a

Figure 6.1   Example of an ecological study.

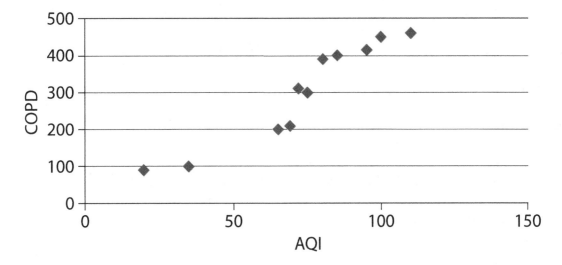

Correlation between the Air Quality Index and Chronic Obstructive Pulmonary Disease

Figure 6.2   Cross-sectional studies capture a snapshot of a population at a specific period in time.

cohort study (see chapter 7) to gather baseline data. They can be used to count the number of cases of a specified condition during a predetermined time period, determine the distribution of factors of interest within a population, and assess the relationship between variables of interest at a given point. Cross-sectional studies provide us with a snapshot of the health status of a population during a particular time period.

Cross-sectional studies are valuable because they provide evidence of the need for analytic epidemiological study. If there is a high prevalence of a particular condition or a consistent association between variables, this may be taken as a sign that further analytic epidemiological studies are warranted. For example, researchers can conduct a quick survey of blood pressure and sodium consumption at the local shopping mall on a busy weekend. The result of this study might indicate that individuals in this area have relatively high blood pressure and consume a high-sodium diet. This would indicate that it might be worthwhile to conduct further research to assess this association more closely.

Often when a cross-sectional study is conducted, an epidemiologist will calculate a prevalence ratio to quantify the prevalence of disease in two groups. Using the high blood pressure example from above, suppose the following 2x2 bale is constructed at the completion of the survey:

	High BP	Normal BP	Total
High salt diet	400	200	600
Low salt diet	100	300	400

N = 1,000

The prevalence of high blood pressure in the whole population is:

= 500/1,000

= 0.50

If we multiply 0.5 by 100, we can say the that high blood pressure is prevalent in 50% of the population.

The prevalence of high blood pressure among those who eat a high salt diet is calculated by dividing the value in the a cell by a+b.

= 400/600
= 0.67

The prevalence of high blood pressure among those who eat a low salt diet is calculated by dividing the c cell by c+d.

= 100/400
= 0.25

The prevalence ratio comparing those with a high salt diet to those with a low salt diet is calculated by dividing the prevalence in the high salt diet group by the prevalence in the low salt diet group. Or:

$$\text{Prevalence ratio} = \frac{a/a+b}{c/c+d}$$

= 0.67/0.25
= 2.68

We interpret this prevalence ratio as follows:

The prevalence of high blood pressure is 2.68 times higher among those who eat a high salt diet compared to those who eat a low salt diet.

If the prevalence ratio equals 1.0, we conclude the prevalence in the two exposure groups are similar. If it is greater than 1.0, we conclude that the prevalence of disease in the group with the exposure is higher than the unexposed group. When the prevalence ratio is less than 1.0, we conclude there is a higher prevalence of disease in the group without the exposure.

There are several weaknesses associated with this type of study design. Researchers are unable to establish the sequence of events in a cross-sectional study. As the variables are all collected at the same time, it is difficult to determine if an exposure actually occurred prior to the outcome; consequently, time-related factors cannot be tested. In our previous example, we found an association between high blood pressure and a high sodium diet, but we cannot say that consuming a high-sodium diet causes high blood pressure, because we do not know which came first. Did the high sodium diet occur first, followed by the high blood pressure? Or was high blood pressure an already existing problem prior to consuming salty snacks and other foods containing sodium? It is logical to assume the consumption of sodium containing food led to the high blood pressure, but we cannot make that assumption with a cross-sectional study design.

Cross-sectional studies are not useful for studying rare conditions because an individual with a rare condition will be difficult to find and may not participate in the study. For example, suppose you are interested in studying maple syrup urine disease, an inherited metabolic disorder that affects how the body processes certain amino acids leading to poor feeding, vomiting, lethargy, and developmental delays. The National Institutes of Health estimate that this disease affects roughly 1 in 185,000 infants worldwide. It is unlikely that you will find this disease during a screening of babies with failure-to-thrive issues in most towns in the United States. Interestingly, the disorder has a higher prevalence in Old Order Mennonite populations, with about 1 in 380 newborns screening positive for the disease. Thus, in a population with a genetic predisposition to a rare disease, it is more likely that the disease will be reported in a cross-sectional study. In general, there are stronger study designs to use to examine rare diseases than cross-sectional studies.

EXERCISES

For each example, determine the following:

- E, the exposure
- D, the disease or outcome
- Study design: Case report, case series, ecological study, cross-sectional study

1. As a part of a hypertension awareness program, volunteers at a weekend health fair in a shopping mall measured the blood pressure and determined smoking habits of 2000 people.

2. The average rate of dental cavities among children in 50 communities was compared with the average level of fluoride in the drinking water of those same communities.

3. Eight women between the ages of 25 and 45 who had undergone breast augmentation with silicone breast implants were reported to have developed severe autoimmune diseases within two to four years after surgery.

4. Twenty individuals who walked barefoot on a gym floor reported the development of plantar warts on the bottoms of their feet.

5. At the local university health fair, 300 people were asked whether they ate salty snack foods and whether they had high blood pressure.

6. The average pollen count and the average rate of sinus infections from several countries were correlated. The results suggest that countries with high pollen counts also had high rates of sinus infections.

7. One person, after undergoing hip replacement surgery, reported increased concentrations of chromium in his blood suggestive of metallosis.

8. A small town in rural Pennsylvania conducted a cross-sectional study to determine the prevalence of depression among community members. It is suspected that depression may be associated with smoking status. Four-hundred individuals participated in the study. It was determined that of the 100 individuals with depression 70 were smokers. Of the 300 individuals without depression, 175 were not smokers.

   a.   Complete the corresponding 2x2 table.

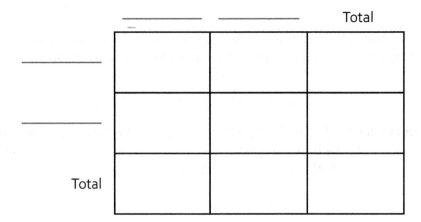

   b.   What is the overall prevalence of depression in the whole population?

   c.   What is the prevalence of depression among the smokers?

   d.   What is the prevalence of depression among the non-smokers?

   e.   Calculate the prevalence ratio comparing smokers to non-smokers.

   f.   How do you interpret the prevalence ratio?

7 ANALYTIC STUDIES—COHORT STUDIES

INTRODUCTION TO ANALYTIC EPIDEMIOLOGY

In descriptive studies, epidemiologists aim to observe the frequency of health related states and provide a means of organizing, summarizing, and quantifying data by person, place, and time. These types of studies are useful in determining disease burden, planning public health programs, and establishing policies to improve health. But what happens when we want to determine what is causing a disease? Or what behaviors will help to reduce the incidence of disease?

To answer such questions, an analytic epidemiological study must be conducted. In this chapter and the two following it, we will describe and discuss the three main types of analytic studies in epidemiology:

- Cohort
- Case-Control
- Experimental

Each design has its own unique characteristics and subsequent strengths and limitations; however, all three share a common goal: to quantify the association between a risk factor/exposure and a specific disease/health outcome.

Table 7.1   Comparison of Descriptive and Analytic Studies

TYPE	DESCRIPTIVE STUDIES	ANALYTIC STUDIES	
		Observational Studies	Experimental Studies
Examples	• Case reports • Ecological Studies • Cross-sectional Studies	• Cohort studies • Case-control studies	• Intervention Trials
Use	Describe disease distributions by person, place, and time	Test a hypothesis that a specific exposure is associated with a specific disease	

Each analytic study design requires the researcher to:

1. Test a hypothesis
2. Calculate a measure of association between disease and exposure.

The null hypothesis (H_0) for all analytic epidemiology studies is that there is no association between the disease and exposure. (REMEMBER: We want to find evidence against H_0. We conduct hypothesis-generated research to determine what disease risk factors exist.) The alternative hypothesis (H_A) is that there is an association between the disease and exposure.

COHORT STUDIES

In this chapter, we will discuss one type of analytic study: **cohort studies**. There are two types of cohort studies—prospective and retrospective. Both are the optimal observational study design for assessing the association between exposure and disease. For a cohort study, individuals must be free of disease at the beginning of the study. Individuals are placed into comparison groups based on their level of exposure and are then followed over time to determine whether disease develops. Both prospective and retrospective cohort studies follow the natural timing of events. Using this study design, incidence of disease among individuals with and without the exposure of

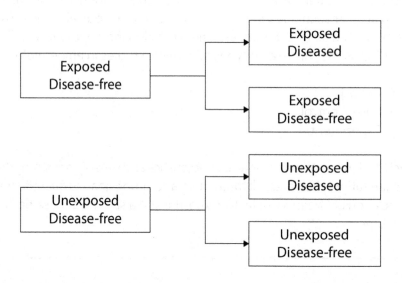

Figure 7.1   Design of Cohort Studies

interest can be determined as individuals are free of disease at the beginning of the study.

Individuals participating in a cohort study must:

1. Be at risk of developing the disease of interest at the beginning of the study;
2. Be recruited into the study according to their exposure status.

> Individuals AT RISK of disease must be:
>
> • Alive at the start of the study
> • Disease-free at the start of the study
> • Capable of developing the disease of interest

The exposed and unexposed groups in a cohort study are followed through time to determine the number in each group that develops disease. Using a 2×2 table, we can see that individuals are assigned to the a+b (exposed) and c+d (unexposed) groups (totals in blue in Figure 7.2) at the beginning of the study. At the end of the study, the number of cases of disease in the exposed group can be compared to the number of cases of disease in the unexposed group.

Because individuals are free of disease at the beginning of a cohort study, we gather incidence data. Using incidence data, we calculate a **relative risk** (RR), which is the measure of association for a cohort study. It quantifies the risk or incidence associated with the exposure.

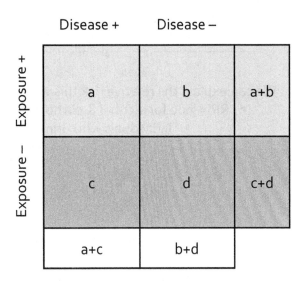

Figure 7.2 2x2 Table for Cohort Studies

If we find the risk of disease is greater in the exposed group compared to the unexposed, the relative risk will be greater than 1.0.

If the risk of disease is the same in both the exposed and unexposed groups, the relative risk will equal 1.0.

If the risk of disease is greater in the unexposed group compared to the exposed group, we say that the relationship between disease and exposure is protective. The value of the relative risk of a protective association between disease and exposure will always be less than 1.0.

> Relative Risk = Risk of disease in exposed divided by risk of disease in unexposed
>
> $$RR = \frac{I_E}{I_U} = \frac{a/a+b}{c/c+d}$$

HOW TO INTERPRET A RELATIVE RISK

We can explain the relative risk this way:
- RR = 3.00 for alcohol & cirrhosis
  - Individuals who drink alcohol are 3 times more likely to develop cirrhosis than individuals who don't drink alcohol
    - risky association
- RR = 0.50 for broccoli & colon cancer
  - People who eat broccoli are 0.5 times less likely to develop colon cancer than people who don't eat broccoli
    - protective association
- RR = 1.00 for coffee consumption & diabetes
  - Individuals who drink coffee are no more likely to develop diabetes than individuals who don't drink coffee
    - no association

Timing of Cohort Studies

There are two types of cohort studies: prospective and retrospective. In a prospective cohort study (also known as a concurrent cohort), the group of participants is recruited in the present/current time and followed into the future to determine disease incidence. Retrospective cohorts, on the other hand, require that the group of participants be assembled in the past, and disease incidence is calculated in the present (see Figure 7.3).

Prospective Cohort Studies

Suppose a researcher wants to determine if oral contraceptive use is associated with an increase in uterine cancer before age 60. An example of a prospective cohort study to answer this question would first involve recruiting two groups of women. These women would then be grouped based on their exposure status: those who use oral contraceptives and those who do not. Both groups—oral contraceptive users and nonusers—would be followed into the future for a specific period of time to determine how many in each group developed uterine cancer before her 60th birthday.

Note: All women participating in this study must be disease-free and at risk of developing the disease at the start of the study. Women with a history of uterine cancer would not be disease-free at the start of the study and could not participate. Similarly, women who have previously undergone removal of their uterus (had a hysterectomy) would not have the target organ of interest.

As these women are followed throughout the study, it is important to minimize the number who are lost to follow-up. Every effort—sending birthday cards, requiring annual check-ups, making quarterly phone calls to stay connected—must be made to ensure that every participant completes the study. Selection bias can occur if there are important differences between women who are and are not lost to follow-up.

The primary advantage of a prospective cohort study is that the temporal relationship

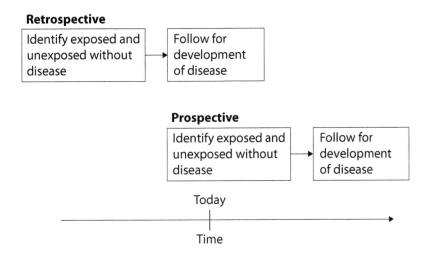

Figure 7.3   Timing of Cohort Studies

between exposure and disease can be established. We know that the exposure occurred before the disease because individuals were disease-free at the beginning. This provides us with incidence data and a relative risk can be calculated. Additionally, exposure status can be defined and measured accurately by the researcher and is not dependent on recall or records. Likewise, the assessment of exposure status of each participant will not be biased by knowledge of the disease of interest because all individuals participating in a prospective cohort study are disease-free at the start of the study.

Prospective cohort studies are commonly used to study rare exposures and when multiple health effects of a single exposure are of interest to the researcher. In our example looking at the association between oral contraceptive use and uterine cancer, the researcher could also follow participants to determine if they developed breast cancer, heart disease, or diabetes. The only requirement is that participants be free of breast cancer, heart disease, and diabetes at the start of the study.

The primary limitation of prospective cohort studies is that they tend to be long and expensive, especially if we are following a disease with a long **latency period**. They also tend to require a large study population. Additionally, prospective cohort studies should not

be used to study rare diseases. If a disease is rare, we would have to enroll a large cohort of exposed and unexposed individuals in order to have enough cases of disease arise during the study period. If enough disease doesn't occur in either or both the exposed or unexposed group, the study will lack **power** and we will not be able to confidently calculate the appropriate measure of association.

Retrospective Cohort Studies

Suppose a researcher wants to determine if exposure to benzene is associated with an increased risk of death from brain cancer. An example of a retrospective cohort study to answer this question would include first obtaining exposure records from factory workers who were potentially exposed to benzene in the past. Using these records, the researcher would have to determine the exposure status of each employee based on the records from the past, establishing benzene-exposed and unexposed groups. Then the researcher would use current death records—in the present—to determine which employees died of brain cancer after working in the factory.

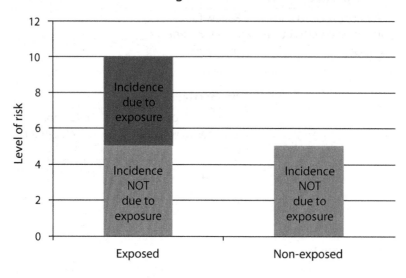

Figure 7.4. Understanding Attributable Risk Percent.

All participants in a retrospective cohort study must:

- Have past exposure data available (possible sources of previous exposure data include occupational health records, medical records, and residential history, among others).
- Be disease-free at the time exposure data was initially collected (remember, this data must be available in the past).
- Be followed from the past to the present to determine disease status.

Like prospective cohort studies, retrospective cohort studies are able to measure risk of disease; the temporal relationship between exposure and disease can be reliably established. Since records from the past are used to establish exposure and disease status is determined in the present, retrospective cohort studies tend to be less expensive and time consuming to conduct compared to prospective cohort studies.

Retrospective cohort studies, like prospective cohort studies, are often used to study rare exposures and multiple health effects. For example, going back to our example research question about the association between benzene and brain cancer, the researcher could also use the retrospective cohort design to study the relationships between benzene and deaths from other types of cancer such as leukemia, lymphoma, and breast cancer, among others.

The primary limitation of retrospective cohort studies is that they rely on previously recorded data from the past to establish the exposure groups. In order to complete a retrospective cohort study, the exposure data must be available in the past and individuals must be followed to the present in order to determine health status. A researcher conducting a retrospective cohort study has no control over the quantity or quality of the exposure data.

ATTRIBUTABLE RISK PERCENT

Looking back at Figure 7.2, recall that cell C in our 2x2 table represents individuals who are not exposed to a risk factor but who develop the disease of interest. These individuals would develop the disease whether they were exposed or not. We call this baseline disease or background risk. To obtain a clearer picture of how many cases of disease might be associated with an exposure, we remove baseline disease.

The attributable risk percent (AR%) tells us the proportion of disease among the exposed that is due to the exposure. AR% is calculated as follows:

$$\frac{Relative\ Risk - 1}{Relative\ Risk} \times 100$$

Attributable risk percent is a measure of potential impact and can be used as a way to quantify the value of a prevention program. It tells how much disease might be avoided if the exposure were eliminated. For instance, we could calculate how much of a reduction in lung cancer we might expect if all smokers stopped smoking. Attributable risk percent is very useful in helping to determine health priorities and policies. A program that has the potential of eliminating a large percentage of cases would have more priority for funding than a program that could potentially eliminate a smaller percentage of cases.

It is important to note that AR% can only be calculated for cohort studies because AR% is based on the relative risk, which in turn is based on incidence data. You can only have incidence data when the study begins with disease free participants, which is one of the criteria for cohort studies.

EXERCISES

The Veterans Administration was interested in assessing whether the use of Agent Orange during the Vietnam War is associated with increased risk of death from lymphoma. Agent Orange is a defoliant used during the Vietnam War. According to the Agency for Toxic Substances and Disease Registry (ATSDR), Agent Orange and other herbicides have been hypothesized to be associated with an increased risk of cancer, among other health concerns. To assess this relationship, researchers reviewed military personnel files to find soldiers who served in Vietnam. Of the 10,000 records that met this criterion, 5425 soldiers were determined to have been exposed to some level of Agent Orange. Of these soldiers, 2175 were exposed to high levels of Agent Orange; the remaining soldiers were determined to have a low exposure to the herbicide. A search of these soldiers' death records revealed that 675 of the soldiers in the high-exposure category had died of lymphoma. In the low-exposure category, 250 soldiers died of lymphoma.

1. Complete the 2×2 table, labeling the exposure and disease boxes with the proper information. Show all your work for the calculations.

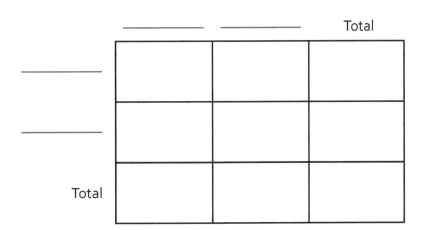

2. What is the exposure and what is the outcome?

3. What type of cohort study is this—prospective or retrospective? How do you know?

4. Calculate the appropriate measure of association between the exposure and the outcome. How would you interpret in words the association you found?

A group of 5000 women in their 30s were identified as having had tuberculosis in their adolescent years. They were then classified into groups according to whether they had a chest x-ray to monitor their disease. 2000 had a chest x-ray and 3000 did not have a chest x-ray. These women where then followed for 40 years to see if those receiving an x-ray were more likely to develop breast cancer. Of those receiving a chest x-ray, 60 were diagnosed with breast cancer. Of those not receiving a chest x-ray, 65 developed breast cancer. Use this data to complete the 2x2 table, being sure to clearly state the exposure and outcome.

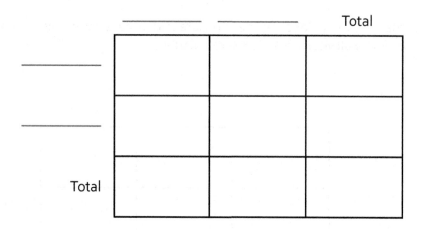

5. Is this a prospective or retrospective cohort study? How do you know? What is the appropriate measure of association?

6. Compute the appropriate measure of association between the exposure and disease and interpret this statistic.

7. Calculate the attributable risk percent and explain it.

8

ANALYTIC STUDIES—CASE-CONTROL STUDIES

A second type of analytic epidemiological study is the case-control study. Also known as case-referent studies, **case-control studies** compare the exposures among individuals with a specific disease to those without the disease. In a case-control study, individuals with the disease of interest (cases) are recruited to participate, as are individuals without the disease (controls). Both groups are assessed to determine exposures in the past.

CONDUCTING A CASE-CONTROL STUDY

Imagine a researcher is interested in determining if exposure to pesticides is associated with bladder cancer. Using a case-control design, the researcher would recruit a group of individuals with a diagnosis of bladder cancer. Similarly, a group of individuals without bladder cancer will be recruited. Both groups would then be asked about their past exposure(s) to pesticides.

Because we know at the beginning of the study who has the disease of interest (a+c cells of our 2×2 table) and who does not (b+d), we are not able to calculate incidence of disease or relative risk.

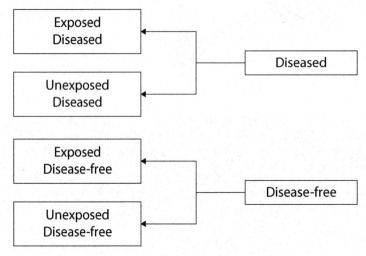

Figure 8.1   Design of Case-Control Studies

The **odds ratio** (OR) is the measure of association calculated for a case-control study. This is it is a ratio comparing the odds of exposure among those with the disease to the odds of exposure among those without the disease. Incidence and relative risk cannot be calculated because we are not watching participants develop disease; no incidence data is collected. Rather, we are asking groups of individuals with the disease and without the disease whether or not they have the exposure of interest.

$$OR = \frac{\frac{a}{c}}{\frac{b}{d}} \text{ or } \frac{ad}{bc}$$

Case-control studies make it possible to study rare diseases, as well as multiple exposures that may be associated with a single disease. For instance, if we go back to the example study at the beginning of this chapter about the association between bladder cancer and pesticides, we can imagine that the researcher

---

HOW TO INTERPRET AN ODDS RATIO

The odds ratio can be explained as follows:
- OR = 3.00 for alcohol & cirrhosis
  - Individuals with cirrhosis (cases) are 3 times more likely to drink than individuals without cirrhosis (controls)
  - Risky association
- OR = 0.50 for broccoli & colon cancer
  - People with colon cancer (cases) are 0.5 times less likely to eat broccoli as people without colon cancer (controls)
  - Protective association
- OR = 1.00 for coffee consumption & diabetes
  - Persons with diabetes (cases) are no more likely to drink coffee than are people who are not diabetic (controls)
  - No association

	Disease +	Disease −	
Exposure +	a	b	a+b
Exposure −	c	d	c+d
	a+c	b+d	

Figure 8.2   2x2 Table for Case-Control Studies

could also investigate the relationships between bladder cancer and smoking, alcohol use, and exposures to dry-cleaning solvents, to name a few. One health effect and a series of potential risk factors are examined.

Case-control studies are often conducted when medical care is sought for the disease of interest. Cases are selected directly from the institution where medical care is being given. When selecting cases to participate in a case-control study, it is recommended that new/recently diagnosed cases are included in the study. Recruiting newly diagnosed cases is beneficial because participants are more likely to reliably recall past exposures if they occurred more recently. Asking someone diagnosed with cancer ten years ago to remember their experiences and exposures prior to diagnosis is more challenging and likely less reliable than someone diagnosed within the past year.

Selecting the non-diseased (known as the control group) is usually the most difficult aspect of designing a case-control study. Controls should be selected from the same source population as the cases. They must be selected independently and without knowledge of each participant's exposure status.

The control group in a case-control study does not need to be representative of the general population. It does need to be representative of the population that gives rise to the cases. Controls, therefore, should be representative of the population that would have become cases if they had gotten the disease.

Types of Controls and Matching

Two types of controls used in case-control studies include:

- Population-Based;
- Hospital.

Population-based controls are the ideal option when conducting a case-control study. However, sometimes selecting a population-based sample is not feasible. Population-based controls are a random sample of individuals without the disease of interest who are selected from the same **source population** as the cases. Population-based controls are often selected through random digit telephone dialing or population registries such as driver's licenses or health insurance records.

Hospital controls are selected from patients seeking care for conditions other than the disease of interest at hospitals or other healthcare facilities where cases are identified. One assumes that controls would be representative of the source population that the cases came from. When using hospital controls, it is advisable to select control patients who have a variety of diagnoses. Additionally, patients with diseases thought to be related to the exposure of interest should be excluded. For example, if we go back to our example about pesticides and bladder cancer, let's assume we decide to select controls that are seeking care at the same hospital as the cases. We know that long-term exposure to pesticides is associated with neurological disorders such as Parkinson's disease. Therefore, we might decide not to include individuals with Parkinson's disease in our control group because the disease for which they are seeking care is known to be associated with the exposure we want to study.

Additionally, certain exposures/risk factors such as smoking, obesity, and older age are overrepresented among hospital patients because these exposures are associated with many diseases. It is therefore recommended that hospital controls not be used when the exposure of interest is associated with a host of different diseases.

Whether hospital or population-based controls are used in a case-control study, it is common to match cases and controls. The purpose of **matching** is to make the case-control groups as similar to each other as possible with respect to one or more variables. Matching is one way of controlling for confounding in the design phase of the study.

There are two ways of matching cases to controls: frequency matching and individual matching. **Frequency matching** requires that controls selected for the study have a similar distribution of a matching variable among the cases. For instance, if we decide to frequency match cases and controls based on gender, first we would recruit all of the cases into the study and determine the ratio of males to females. Then we would seek to recruit a group of controls with a similar gender ratio.

Individual matching requires the researcher to match each control to a particular case with respect to the matching variable(s). If we wanted to use individual matching in our bladder cancer and pesticides example, we would first determine the matching variable. In this case, it is gender. If a male case is recruited, the researcher must then go and recruit a male control. Often, when individual matching is used, we match more than one control (up to four) to each case. Doing so increases both the power and sample size of the study. Matching will be discussed further in Chapter 13 Confounding and Effect Modification.

The case-control study design is most often used (and best suited) to investigate the effects of multiple exposures on a single health effect. It is also the ideal design to study rare diseases. The process of conducting a case-control study requires that cases and controls be identified first; we are not waiting for the disease to arise within a population that was disease-free at the start of a study. If a disease is rare, the source population for the study can be expanded in order to recruit the appropriate number of cases to give the study enough **power**. In comparison to cohort studies, the case-control study design usually will result in shorter (and often less expensive) studies.

However, case-control studies do have limitations. Since individuals are recruited into case-control based on their health status, the event of interest (development of disease) has already occurred. Therefore, neither incidence nor relative risk can be calculated and that is why we use an odds ratio. Case-control studies are also prone to recall and selection bias. Despite these limitations, case-control studies are used frequently by epidemiologists. They are used to identify genes associated with cancer, sources of food-borne outbreaks, and environmental factors that exacerbate asthma. Epidemiologists recognize both the strengths and limitations of case-control studies and discuss the results of studies in a way that addresses both.

EXERCISES

The use of cellular phones has been associated with the development of brain cancer, as evidenced in several studies; however, other studies have shown no such association. To further investigate this association, researchers recruited participants based on whether or not they had a specific type of brain tumor. A total of 171 individuals had brain tumors and 492 did not. Of the individuals who had a brain tumor, 75 used their cell phones for more than 200 minutes every day. Of the individuals who did not have a brain tumor, 292 used their cell phones for 200 or less minutes every day.

    Complete the 2×2 table, labeling the exposure and disease boxes with the proper information. Show all your work for the calculations.

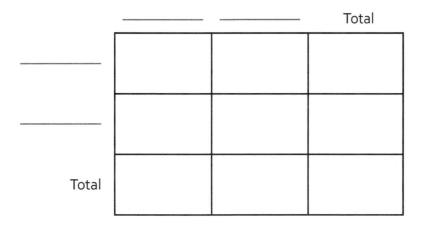

1. What type of study design is this? How do you know?

2. What is the appropriate measure of association to use for this study design? What is the formula for this measure of association?

3. Calculate the proper measure of association. Interpret your results in words, indicating whether it is risky, protective, or unassociated.

9 ANALYTIC STUDIES— EXPERIMENTAL STUDIES

The final type of analytic epidemiological study is the experimental study. This category of studies can be further broken down into three subtypes:

- Clinical trials: Used to evaluate a new treatment for an ill patient—for example, a new type of chemotherapy for cancer patients.
- Field trials: Employed to evaluate interventions for disease prevention—for example, testing a new vaccine aimed at preventing disease.
- Community trials: Used to evaluate community-wide interventions—for example, comparing the oral health of two communities: one with fluoride in the drinking water and one without fluoride.

All three types of **experimental studies** require that participants be placed into a treatment or control/placebo group and followed over time. As we discuss the specifics of experimental studies on the

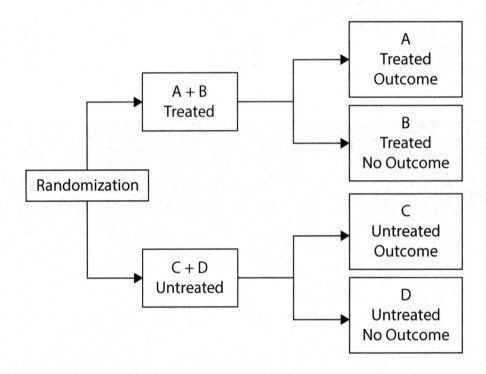

Figure 9.1   Design of a Randomized Control Trial

pages that follow, we will refer to them as **randomized control trials** (RCTs).

According to Leon Goridis, prominent epidemiologist and author, "The randomized trial is considered the ideal design for evaluating both the effectiveness and the side effects of new forms of intervention." It is the gold standard for epidemiological study designs.

In an RCT, treated and untreated participants are followed over time to determine whether they experience the outcome.

Individuals are assigned to their treatment group through a process called randomization. **Randomization** is the process by which all subjects have an equal probability of being assigned to either the intervention or control group. The process of randomization should be unbiased, unpredictable, and tamper-proof.

The goals of randomization are to:
- Remove all the potential for bias in the allocation of subjects to the treatment groups;
- Make sure the treatment groups are comparable with respect to known and

unknown factors (potential confounders).

Often (when possible), participants in an RCT will be **blinded** (unaware of) to their treatment assignment until after the trial is complete. Single blind means that the participant doesn't know to which group he or she is assigned. If both the participant and his or her physician are unaware of the treatment group assignment, we say the study is double blind. If the participant, physician, and data analyst are unaware of which treatment group has been assigned, we say the study is triple blind.

In order to conduct an RCT, the following criteria must be met:

- Have preliminary evidence of treatment/interventions, safety, and efficacy.
- Know enough about treatment/interventions to know which outcomes to assess.
- Must be conducted before the treatment becomes part of standard medical practice.

$$\text{Efficacy} = \frac{Rate_{placebo} - Rate_{treatment}}{Rate_{placebo}} \times 100 \quad or \quad \frac{c/c+d - a/a+b}{c/c+d} \times 100$$

Figure 9.3   Calculation of Efficacy

In epidemiology, we say that an RCT can be conducted if—and only if—there is a state of equipoise. According to the Dictionary of Epidemiology, **equipoise** is:

> "A state of genuine uncertainty about the benefits or harms that may result from each of the two or more regimens. A state of equipoise is an indication for a randomized controlled trial because there are no ethical concerns about one regime being better for a particular patient."

Because participants in both the intervention and control arms of the study are followed across time to determine if a specific health outcome of interest occurs, we can determine if the intervention is more effective than the control (often the standard of care) or **placebo**. The measure of association most often used to quantify the difference in effect between groups is efficacy. **Efficacy** measures a reduction in risk.

Specifically, efficacy is the proportion of individuals in the control group who experience the outcome of interest (which is not desired), who could have been expected to have a favorable outcome had they been in the intervention group instead. We can use our 2x2 table to calculate the rate of the outcome in both the treated and untreated/placebo groups.

Suppose we have conducted a randomized control trial to determine if a hand-washing education program will result in fewer cases of influenza in a nursing home. Participants were randomized into two groups: the treatment group received a comprehensive education in hand-washing; the placebo group received no education. The results of the study are summarized in the following 2x2 table:

	Influenza	No influenza	
Hand-washing education	250	750	1,000
No handwashing education	500	500	1,000

N = 2,000

We calculate the efficacy of the hand-washing education program as follows:

Rate placebo = c/c+d
= 500/1000
= 0.50

Rate treatment = a/a+b
= 250/1000
= 0.25

Efficacy of hand-washing education = $\dfrac{0.5 - 0.25}{0.5}$
= 0.25/0.5
= 0.50

We interpret this finding by stating either:

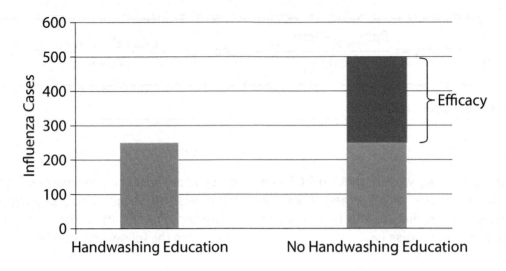

Figure 9.2 Illustration of Efficacy

The efficacy of the hand-washing education program at preventing influenza in our population is 50%.

Or we can say, 50% of the influenza infections would have been avoided in the control group if they had received the hand-washing education.

EXERCISES

A study was conducted to determine whether the drug AZT was effective in preventing serious opportunistic infection among individuals infected with HIV. Individuals who were HIV-positive were randomly selected to either receive AZT treatment or a placebo (nondrug) treatment. Among the 145 individuals who received AZT, 24 developed serious opportunistic infections. Among the 137 individuals who were not treated with AZT, 45 developed serious opportunistic infections.

1. What type of study is this? How do you know? Be specific.

Complete the 2×2 table for these data. Be sure to identify what the outcome and no outcome groups are, as well as the treatment and not treated groups.

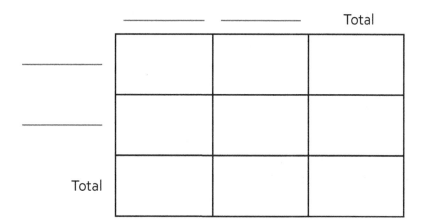

2. What is the measure of association for this study type? Calculate the association between AZT treatment and the development of opportunistic infections.

3. Explain what this measure of association means.

First, read the article:

Risk and Benefits of Estrogen Plus Progestin in Healthy Postmenopausal Women: Principal Results from the Women's Health Initiative Randomized Control Trial, by the Writing Group for the Women's Health Initiative Investigators, in the *Journal of the American Medical Association* 288(3): 361.

READING THE EPIDEMIOLOGICAL LITERATURE

In order to read, understand, and evaluate peer-reviewed journal articles about epidemiological studies, you must train yourself to think like an epidemiologist.

For each study you read, try to answer the questions:

Who? What? Where? When? Why? and How?

Specifically, you should identify and critically analyze each study's:

1. Research hypothesis/question
2. Study design
3. Independent and dependent variables
4. Data collection methods
5. Methods of analysis
6. Potential confounders and if/how they were controlled
7. Possible sources of bias
8. Interpretation and utility of results

And then answer the following questions:

4. What was the objective(s) of the trial?

5. Who made up the study population?

6. What were the treatment/control groups?

7. How were the participants randomized?

8. What is blinding?

9. Who was blinded in this study?

10. What are the advantages of blinding?

11. How would you summarize the results of the study?

12. What are the implications of the results?

10

PROGNOSTIC STUDIES AND SURVIVAL ANALYSIS

Up to this point, we have discussed how to design epidemiological studies to determine what risk factors or exposures have caused disease. What if we want to answer questions about the future course of a disease once it has been diagnosed? How do we use epidemiology to answer questions such as:

- How long am I expected to survive now that I've been diagnosed with a specific disease?
- What type of therapy is appropriate for treatment based on my diagnosis?
- How long can I expect to live following successful treatment of a disease?

PROGNOSTIC STUDIES

Prognostic studies are conducted in order to make a prediction of the future course of a disease. Studies of prognosis usually focus on outcomes important to the patient such as death, complications, pain and suffering, quality of life, and remission. Both positive and negative outcomes are studied in prognostic analyses.

Several of the measures of association used in prognostic studies, such as case-fatality and death rates, are discussed in detail in Chapter 5 Mortality. When discussing these measures in terms of prognosis, we state:

- **Case fatality** is the proportion of patients with a disease who die from it.
- **Death rate** is the total number of deaths observed divided by the total amount of person-time identified among the patients participating in the study.

Because the calculation of a death rate assumes a constant death rate over time, we often calculate a five-year survival. This measure calculates the proportion of patients who survive five years from a defined point in the course of the disease. This defined point could be the time of diagnosis or the completion of treatment. Calculating the five-year survival rate can only be done after all of the patients in the study have been followed for at least five years.

A more efficient way to use all the data on a cohort of patients followed over time is the life table approach (also called the actuarial method). **Life table analysis** results in a calculation of the probability of surviving to a point estimated from a cumulative probability of surviving each of the previous time intervals. Cumulative probabilities are the product of period-specific probabilities.

Suppose we gather the following data from a cohort of patients with pancreatic cancer:

Table 10.1   Life Table Approach Pancreatic Cancer Example

Year of Diagnosis	Number	ALIVE AT END OF YEAR		
		Year 1	Year 2	Year 3
2011	75	60	50	45
2012	65	55	30	
2013	40	35		
TOTAL	180	150	80	45

Using the table above, we can answer numerous questions related to prognosis:

1. What is the probability of surviving the first year?

   $= 150/180$
   $= 0.83$ of 83%

2. What is the probability of surviving the second year, given that the patient survived the first year?

   $= 80/(150-35)$
   $= 0.69$ or 69%

In this instance, we must subtract the 35 individuals because those patients were only observed for one year.

3. What is the probability of surviving the third year, given the patient survived the second year?

   $= 45/(80-30)$
   $= 0.90$ or 90%

Again, we subtract 30 because those patients were only observed for two years.

4. What is the probability of surviving all three years?

   $=$ product of the probability of surviving the first, second, and third years
   $= (0.83)*(0.69)*(0.90)$
   $= 0.52$ or 52%

We make four major assumptions when using the life table approach:

- We assume that loss to follow-up of patients occurs at the midpoint of the interval. Therefore, we can subtract half of those lost to follow-up during the interval from the total number alive at the start to produce an adjusted number at risk.

The number at risk = # alive at the start − (# lost)/2

- We assume individuals lost to follow-up are similar to those not lost.
- We assume the survival rate is unchanging.
- We assume that patients enrolled at different points are similar.

Other ways of calculating survival such as the Kaplan-Meier method and Cox proportional hazards models require that the researcher know the exact timing of events for each study participant.

EXERCISES

To better understand the life table analysis process, try the example in Table 10.2. Remember the adjusted number at risk = the number alive at the start of the interval—half of those lost to follow-up; the proportion who died in the interval is calculated by dividing the number who died in the interval by the adjusted number at risk; the proportion who did not die in the interval is calculated by subtracting the proportion who did die by 1; and the cumulative probability of surviving to the end of the interval is determined by multiplying all of the previous cumulative probabilities together.

Table 10.2   Life Table Analysis Example

TIME INTERVAL	ALIVE AT START OF INTERVAL	DEAD	LOST TO FOLLOW-UP	ADJUSTED # AT RISK	PROP WHO DIED IN INTERVAL	PROP WHO DID NOT DIE IN INTERVAL	CUM P(SURV) AT END OF INTERVAL
1	50	5	3				
2	40	4	0				
3	30	3	2				
4	20	2	1				
5	0						

11 ETHICS IN EPIDEMIOLOGICAL RESEARCH

Conducting research with human participants (sometimes called subjects) is beneficial and necessary, as well as risky. Ethical treatment of each participant and protection of his or her information must be the priority of any epidemiological study.

The history of research on human participants is full of examples of the unethical treatment of human beings. During World War II, the Nazi government in Germany conducted research on humans. Most of these studies involved individuals who were forced or coerced to participate. As a result, the Nuremberg Code, written following the end of the war, states in its first principle that any individual participating in a research project must provide consent to take part in it. According to the code, if an individual does not consent, then he or she is not part of the study. Voluntary informed consent from each potential study participant is absolutely essential. For this reason, we propose using the term **research participant** instead of *subject*. The term participant conveys that the individual involved in an epidemiological study is voluntarily participating in the research; he/she is aware of his/her participation and is willing to take part in the study in order to further public health research.

Following the establishment of the Nuremburg Code, a series of unethical studies—the Tuskegee Syphilis Study, Milgram's experiment in obedience, the Willowbrook Study, and Zimbardo's Stanford prison experiment to list but a few—spurred the federal government to look at the practices of human participant researchers and establish federal requirements regarding the ethical treatment of humans participating in such research studies.

The resulting code is outlined in the Belmont Report, which was published in 1979. The Belmont Report includes three guiding principles for conducting ethical research with human participants:

- Respect for persons;
- Beneficence;
- Justice.

RESPECT FOR PERSONS

Individuals participating in research should be treated as autonomous agents. Persons with diminished autonomy—the elderly, children, unborn fetuses, prisoners, and the developmentally delayed—are entitled to protection. Researchers who want to study such vulnerable populations are held to higher standards of ethical conduct.

Individuals must agree to participate in a study voluntarily. They must also be given adequate information to make an informed decision about whether or not to participate. We say that there is **informed consent** among all participants.

Informed consent must include the following:

- Research procedures and purpose;
- Risks and anticipated benefits;
- Alternative procedures;
- A chance to ask questions or withdraw.

If the study includes children under 18 years of age, typically, the child's parent or guardian will be responsible for giving the informed consent to participate. Since children cannot legally give their consent, they are asked if they would like to participate, a process called **assent**. If a child refuses to participate—even after a parent has provided informed consent to participate—he or she cannot be required to assent.

BENEFICENCE

Researchers conducting human research have an obligation to each participant to:

- Do no harm;
- Maximize possible benefits;
- Minimize possible harms.

Risks and benefits may affect the participants, their families, and society at large. When considering risks, we should weigh both the probability harm will occur and consider how severe the harm might be. Benefits should outweigh risks and may accrue to society at large. A balance must be struck between benefits and harms. We also need to think about what could be lost if the research is not conducted.

JUSTICE

A researcher must be fair in selecting participants. All participants in a study should be treated the same. Additionally, the selection of participants should be scrutinized to determine why some groups are systematically selected. Often, individuals are selected to participate because the researcher has easy access to the group; the group then is in a compromised position and can easily be manipulated into participating.

All researchers should select participants for reasons directly related to their research question and hypothesis. Persons from groups unlikely to benefit from the research should not be included in the study population. For instance, we would not conduct research with a group of prisoners just because they are readily available and unlikely to feel free to object to participating. We could, however, have prisoners participate in research if the study's results would be directly applicable to improving prison life or addressed a health condition that specifically affected incarcerated people. Additionally, all publicly funded research must ensure that the benefits of the study are available to all participants, including those who are unable to afford them.

INSTITUTIONAL REVIEW BOARD (IRB)

Federal regulations require all institutions, including colleges and universities, to have an **Institutional Review Board** (IRB) in place to assure that the rights and welfare of all human research participants are adequate. An IRB must approve all research projects involving human participants. Written approval of a project must be received from the IRB before a research project can begin.

Most academic institutions also require that each individual participating as a researcher on a project involving human research understands the laws and requirements for conducting ethical research. To ensure this, most institutions require researchers, including students, to complete the online Collaborative Institutional Training Initiative (CITIprogram. org). The CITI program strives to "promote the public's trust in the research enterprise by providing high quality, peer reviewed, web based, research education materials to enhance the integrity and professionalism of investigators and staff" conducting research with human participants.

12 ERROR AND BIAS

According to the *Dictionary of Epidemiology*, bias is:

"Any effect at any stage of an investigation or inference tending to provide results that depart systematically from the true values."

In simpler terms, **bias** is a systematic error in the design or conduct of the study. Biased results will be different from the "true" results and cause a lack of internal validity in the study.

Internal validity is the extent to which the results of a study accurately reflect the true situation in the study population.

If a study lacks internal validity, it will not have external validity. **External validity** is the extent to which the results of the study are applicable to other populations.

There is no way to control for bias, as we do with confounding (see Chapter 13). Our role as epidemiologists is to take the necessary steps to ensure bias is not introduced into our studies. If we do find bias in a study, it is our responsibility to identify and explain it, as well as discuss the impacts it may have had on our results.

Figure 12.1   Relationship between Bias and Validity

TYPES OF BIAS

There is an extensive list of specific types of bias in epidemiology—assembly, Berkson's detecion, lead-time, immortal time, interviewer, observer, and susceptibility bias, among others. Here we will focus on the two most common types of bias in epidemiology:

- Selection bias;
- Information bias.

SELECTION BIAS

Selection bias results from differences in either:
- The characteristics of those who are selected to participate and those who are not; or
- The characteristics of groups within the study: specifically, the association between the exposure and disease differs for those selected into the study and those who are not.

For selection bias to arise, the factor impacting the selection of subjects must be associated with both the exposure and disease. Examples of selection bias include:

Loss to follow-up: This is especially common in prospective cohort studies. If the group of participants lost to follow-up over the course of the study is systematically different from those who complete the study, bias is introduced. For example, if we are conducting a 30 year long prospective cohort study to study the incidence of cancer, participants could be lost to follow-up if they move away from the research center, die, stop participating due to ill health, family responsibilities, or boredom.

Exclusion: Establishing different eligibility rules for recruitment of the comparison groups in a study (i.e., cases and controls) will result in biased results. For instance, if one would decide to exclude all cases who are smokers from a case-control study, but allow smokers to be part of the control group.

Sampling bias: If a **non-random sampling strategy** is used to recruit participants, a systematic error in who was selected to participate may occur. For example, if a researcher decides to conduct a voluntary internet-based cross-sectional study in a small community, sampling bias could occur if individuals in lower socio-economic groups do not have access to the internet. The sample collected for the study would include a biased, unrepresentative sample of the population because it was made up of only those who could afford internet access.

Response bias: A systematic error due to differences in the characteristics between participants. For example, if participants are too sick or already dead, they will not be able to participate, and information about the disease and exposure being studied will not be gathered.

INFORMATION BIAS

Information bias is defined as an error in the classification of participants with respect to disease or exposure status. The two most common types of information bias are misclassification and recall bias.

Misclassification occurs when the disease or exposure status of participants is categorized incorrectly. There are four types of misclassification.

- Non-differential misclassification of exposure;
- Non-differential misclassification of disease;
- Differential misclassification of exposure;
- Differential misclassification of disease.

Non-differential misclassification of exposure occurs when a systematic error results in misclassifying the exposure independent of the disease status. For example, a prospective cohort study focuses on coffee consumption and the risk of migraines. Assume individuals are likely to underreport their coffee consumption. However, individuals who developed migraines were no more likely to underreport coffee consumption than those who did not develop migraines.

Non-differential misclassification of disease happens when a systematic error results in misclassifying the disease independent of the exposure status. For instance, a prospective cohort study is conducted to determine the association between employment in academia and obesity, measured through self-reported height and weight. In general, people tend to underreport how much they weigh and over-report height. If the degree of underreporting weight and over-reporting height is similar for those in academia and those not, we should report a non-differential misclassification of disease.

Differential misclassification of exposure occurs when the misclassification of exposure is not independent of disease status. For example, suppose a case-control study is conducted to determine if taking Tylenol is associated with liver failure. Assume that Tylenol use is underreported in medical charts of patients. If we find that underreporting is more common among those with liver failure compared to those without, we would have differential misclassification of exposure.

Differential misclassification of disease occurs when a systematic error causes the misclassification of disease that is not independent of the exposure status. For example, imagine conducting a study about the association of gym membership and obesity. Participants are classified as obese—a case—if their self-reported BMI is greater than 25. If they self-report a value less than 25, they will be classified as a control (with a healthy weight). If all participants who attend the gym accurately report their height and weight (producing an accurate BMI) and all those who do not attend the gym underreport their weight and over-report their height (leading to an underestimation of BMI), we would have differential misclassification of disease.

Recall bias is a specific type of differential misclassification of exposure. It is most likely to occur in case-control studies. Recall bias occurs when cases (individuals with the disease), remember and report risk factors in a different (more specific) way than the controls. Imagine conducting a case-control study of new mothers soon after the birth of their children. Case mothers will be those with newborn babies with major birth defects and controls will be mothers with healthy newborns. If both cases and controls are given a questionnaire asking them about prenatal environmental exposures, it is possible that the case group will recall their exposures in a systematically different way than the controls.

The mothers of children with a birth defect may respond in more detail to determine what caused the adverse birth outcome. This is different from the way that a mom with a healthy baby would complete the questionnaire.

HOW DO WE ADDRESS BIAS?

The simple answer is to anticipate how and when bias may enter a study. Specifically, we should use the appropriate study design to minimize bias based on our research question and target population. We should also try to maximize response and follow-up rates. To minimize information bias, we should use data collection tools that have been validated and pretested. In addition, we should use the same data collection method(s) for all participants in the study. Blinding the research staff to each participant's disease and exposure status is one way of avoiding differential misclassification.

If bias cannot be avoided or minimized, it is our responsibility as epidemiologists to identify the bias in our study and discuss the impact it may have had on our results. When discussing the impact on results, we start at the null effect. By definition, the **null effect** of a measure of association is the value that indicates no association between exposure and disease: a relative risk or odds ratio equal to 1. Based on the type of bias, we would say bias resulted in either:

- Bias toward the null;
- Bias away from the null;
- Switchover.

Bias toward the null (the null would be equivalent to the RR or OR = 1) is when the observed effect is weaker than the true effect. **Bias away from the null**, on the other hand, occurs when the observed effect is stronger than the true effect. **Switchover** occurs when the observed and true effect are on opposite side of the null value.

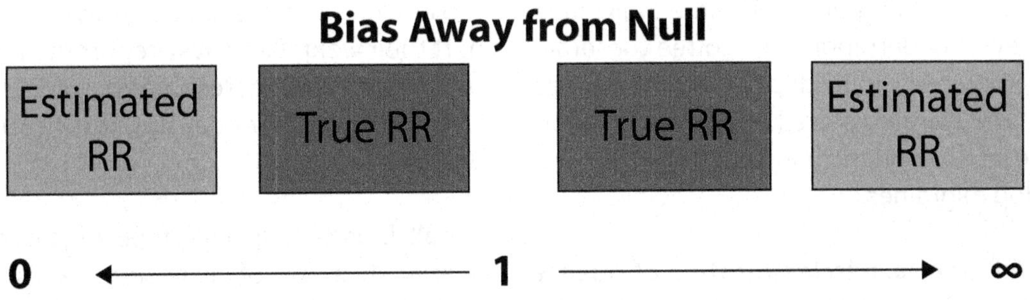

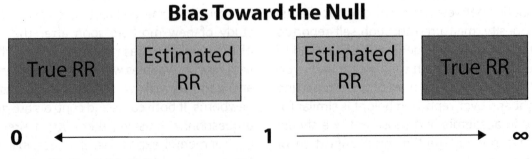

Figure 12.2   Bias Away from Null

Figure 12.3   Bias Toward the Null

EXERCISES

1. Complete the chart below. Indicate the direction of the bias in each situation.

TRUE RR	OBSERVED RR	OBSERVED RR IS BIASED. . .
5.0	3.0	
0.4	0.6	
0.75	3.0	
1.75	4.0	
0.75	0.2	

For each situation below, select the type of bias (if any) described. And provide a brief explanation (one or two sentences) for each choice.

2. A survey of prior exposure to sexually transmitted disease based on self-reporting.
   A. Recall bias
   B. Selection bias
   C. Information bias

3. A study of coronary artery disease, where disease is measured by asking patients the question: "Has your doctor ever told you that you have coronary artery disease?"
   A. Selection bias
   B. Information bias
   C. No bias

4. A study on physical disabilities among the elderly based on a sample of senior citizens attending dance lessons.
    A. Selection bias
    B. Recall bias
    C. No bias

5. It has been suggested that physicians may examine women who use oral contraceptives more often than women who do not use them. A study of these women finds that there is a strong association between oral contraceptive use and blood clots.
    A. Recall bias
    B. Selection bias
    C. No bias

6. A retrospective cohort study used occupational health records to determine the exposure status of participants. Exposure to organochlorine pesticides was defined as anyone working at the company during the summer months from 1965–1970. Everyone who worked at the company during this time period (the six summers) was considered exposed to the pesticides, regardless of the type of work they performed for the company.
    A. Information bias
    B. Recall bias
    C. No bias

13 CONFOUNDING AND EFFECT MODIFICATION

In addition to studying diseases and risk factors, epidemiologists are also interested in studying confounding variables. **Confounding** is a distortion of an exposure-disease association caused by the association of another factor with both the diseases and exposure. This other factor is a **confounder**: a variable that confuses the relationship between exposure and disease.

It is important to study confounders because they interfere with our search for causal associations between exposures and disease. If we do not account for them in our research, policy and programmatic recommendations to reduce disease occurrence will not be effective. For example, imagine conducting a case-control study to investigate the effects of alcohol consumption on the development of lung cancer. It is possible that smoking cigarettes could be confounding the relationship between alcohol and lung cancer. If we conclude that alcohol is the cause of increased lung cancer, we might propose a strategy to reduce alcohol consumption. However, a reduction in alcohol consumption will not lead to a decrease in lung cancer if the association is confounded by smoking.

In order for confounding to occur, the other factor/variable (also called the potential confounder) must be a risk factor for the disease AND be distributed differently among the exposed

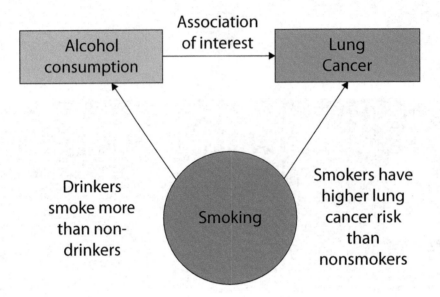

Figure 13.1   Confounding Example—Lung Cancer, Alcohol, and Smoking

DIFFERENTIATING BIAS
AND CONFOUNDING

BIAS
Associations are invalid,
not real because of systematic error

CONFOUNDING
Associations are real, but not causal

and unexposed. Let's look at the smoking/ alcohol consumption/lung cancer example again (Figure 13.1).

By definition, smoking in this example is a potential confounder. Smoking is a known risk factor for lung cancer, and there is evidence that individuals who drink alcohol are more likely to smoke cigarettes.

Once a variable is identified as a potential confounder, we can control for it in our study. If we control for a variable and it changes the estimated effect of the exposure on the disease, it is a confounder.

CONTROLLING FOR CONFOUNDING

Controlling for confounding can be done at both the design and analysis phase of a study. In the design phase, researchers can control for confounding by:

- Randomization;
- Restricting the study to certain groups;
- Matching.

As discussed in the chapter on experimental studies, randomization is a process of assigning individuals to treatment groups in order to make the groups as similar as possible. By making the groups similar, we are distributing the potential confounding variables, those known and unknown, equally among the treatment groups. Imagine randomizing a group of college-aged students into two groups in order to determine if a hand-washing education program reduced the incidence of influenza. If students are randomized into the groups, the process should distribute the potential confounders among the two groups. The groups should be equivalent with respect to potential confounding variables, such as gender, graduation year, major, and other health effects.

Restriction is used when a researcher wants to remove a potential confounder from the study. The researcher will restrict the study population to a particular group or groups that do not have the potential confounder. For example, we know that smoking cigarettes is a causal factor for many negative health outcomes—cancer and heart disease, among others. Smoking is also associated with a host of behaviors—drinking alcohol and lack of physical activity—that are risk factors for cancer, among other diseases. Suppose we were interested in studying the association between pesticide exposure and bladder cancer. Smoking is a potential confounder in this situation. We could restrict our entire study population to nonsmokers to ensure that smoking was not confounding the relationship between pesticides and bladder cancer. Restriction is easily done and understood. It should be noted, however, that if the study population is restricted, external validity will be limited. The study results will not be generalizable to the group that was not included in the study. In the pesticides example, we would not be able to say anything about the risk of bladder cancer associated with pesticide exposure among the smokers.

As discussed in the chapter on case-control studies, matching (often cases to controls) is another way of controlling for confounders at the design stage of a study. The process of matching makes the groups similar with respect to potential confounders. By matching on these factors, we are controlling for confounders. Let's go back to our research question about pesticide exposure and bladder cancer. Suppose we didn't want to restrict the study to all non-smokers. We could match our cases and controls based on their smoking status, thus assuring that an even number of smokers and non-smokers are in both the case and control groups.

The key advantage of matching cases to controls for confounding is that we are assured that the study groups by design are comparable with respect to each matched confounder. There is, however, a set of disadvantages associated with matching, including:

- Increased cost and labor associated with the process;
- Inability to include individuals who are not matched;
- Inability to make conclusions about the relationship(s) between the matched variable and the disease and exposure of interest.

Controlling for confounding can also be done at the analysis phase of a study through:

- Standardization;
- Stratification;
- Multivariable analysis.

As discussed previously in Chapter 5, standardization is the process by which we apply a standard to the study groups to make the underlying population structures similar. We often use age-standardized mortality rates to compare the death rates in geographic areas with different underlying population structures. Similarly, we use age-standardized mortality rates to compare the death rate during different periods of time.

Stratification is the process by which we separate our 2×2 table by the different categories of the confounding variables. For example, imagine we are investigating the relationship between exposure to methylmercury (Methyl Hg) and leukemia. From our data, the following 2×2 table is constructed:

	Leukemia +	Leukemia −	
Methyl Hg +	400	200	600
Methyl Hg −	200	200	400
	600	400	100

If these data were from a case-control study, we would calculate an odds ratio.

Odds Ratio = ad/bc
Odds Ratio = (400*200)/(200*200) = 2.0

This is the crude odds ratio. We would interpret it by saying:

Individuals with leukemia are two times more to have been exposed to methylmercury compared to those without leukemia.

As part of this study, we believe that exposure to benzene may be a potential confounder. The epidemiological literature tells us that benzene is a risk factor for leukemia, and there is evidence that individuals with occupational methylmercury exposures are more likely to be exposed to benzene. Therefore, the potential confounder is associated with the disease and exposure (Figure 13.2).

We can use the process of stratification to determine if benzene is a confounder and control for it if necessary. **Stratification** simply means to separate into groups, in this case based on the confounder. To do this, we construct two 2×2 tables: the first for those with benzene exposure, and the second for those without benzene exposure.

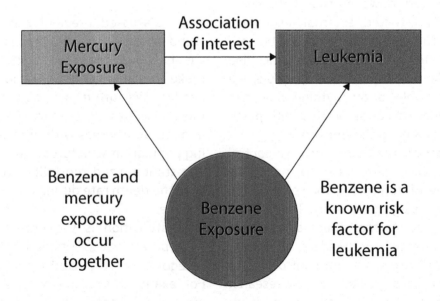

Figure 13.2   Confounding Example: Methylmercury, Benzene, and Leukemia

Benzene Exposed Group:

	Leukemia +	Leukemia −
Methyl Hg +	200	100
Methyl Hg −	100	100

Group Not Exposed to Benzene:

	Leukemia +	Leukemia −
Methyl Hg +	200	100
Methyl Hg −	100	100

The odds ratio for the two benzene exposure groups, or strata, both equal 2.0. Therefore, we conclude that the stratum-specific odds ratios are the same. Our next step is to compare the stratum-specific odds ratio to the crude odds ratio. In this example, we see that the crude and stratum-specific odds ratios are the same. This tells us that the effect of benzene exposure is not confounding the relationship between methylmercury exposure and leukemia. Benzene is not a confounder, and we do not need to adjust for it in the analysis.

Suppose, however, that we conducted the same study among a different population and found the following results:

	Leukemia +	Leukemia −
Methyl Hg +	400	430
Methyl Hg −	380	500

The odds ratio = (400*500)/(380*430)
= 1.22

We conclude that individuals with leukemia are 1.22 times more likely to have been exposed to methylmercury compared to those without leukemia. Again, we believe that exposure to benzene may confound this association. Therefore, we stratify the data by benzene exposure as follows:

Benzene Exposed Group:

	Leukemia +	Leukemia −
Methyl Hg +	250	180
Methyl Hg −	160	120

Group Not Exposed to Benzene:

	Leukemia +	Leukemia −
Methyl Hg +	150	250
Methyl Hg −	220	380

The odds ratios for the two strata based on benzene exposure both equal 1.04. Since they are equal, our next step is to compare them to the crude odds ratio. In this instance, we find that the crude and stratum-specific odds ratios are not equal. Therefore, we conclude that benzene is a confounder. It confuses/confounds the relationship between methylmercury and leukemia. We cannot report the crude odds ratio and must report the stratum-specific value (OR_{ss} = 1.04). Specifically, we state:

> Individuals with leukemia are 1.04 times more likely to have been exposed to methylmercury than those without leukemia, while controlling for benzene exposure.

Through stratification, we can determine if a third variable is a confounder if:

1. The stratum-specific odds ratios are the same;
2. The stratum-specific odds ratios are different from the crude.

What happens when the stratum-specific odds ratios are different?

Assume that in addition to suspecting that benzene exposure confounded the association between leukemia and methylmercury, we also suspect that smoking may be a confounder. Our first step then is to calculate the crude odds ratio between leukemia and methylmercury exposure.

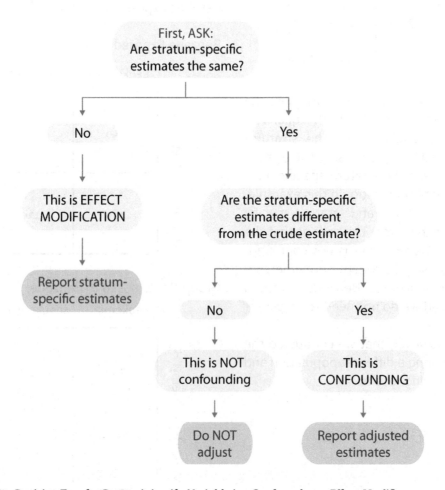

Figure 13.3   Decision Tree for Determining if a Variable is a Confounder or Effect Modifier

	Leukemia +	Leukemia −
Methyl Hg +	400	430
Methyl Hg −	380	500

Now we stratify by smoking status.

Smoking Group:

	Leukemia +	Leukemia −
Methyl Hg +	300	200
Methyl Hg −	300	400

Nonsmoking Group:

	Leukemia +	Leukemia −
Methyl Hg +	100	230
Methyl Hg −	80	100

OR for smokers = (300*400)/(300*200)
$$= 2.0$$
OR for nonsmokers = (100*100)/(80*230)
$$= 0.54$$

The odds ratios for the two strata are not equal. In fact, the odds ratios for smokers is hazardous, and for nonsmokers it is protective. Since the stratum-specific odds ratios are not equal, we cannot compare them to the crude. Instead, we conclude that smoking is an effect modifier.

Effect modifiers, also known as interactions, occur when the stratum-specific measures of association are not uniform. If a third variable is an effect modifier, then the association between exposure and disease will be different at different levels of that variable. When we find effect modification, we report each stratum-specific measure of association. In our example above, we would state:

Among the smokers, those with leukemia are 2 times more likely to have been exposed to methylmercury compared to those without leukemia. Among the non-smokers, those with leukemia are 0.54 times less likely to have been exposed to methylmercury compared those without leukemia.

EXERCISES

A prospective cohort study of the effect of milk consumption and the incidence of bad breath was conducted. The overall results of the study are shown in the table below:

Incidence of Bad Breath by Milk Consumption

GROUP	INCIDENCE OF BAD BREATH
Consume milk	0.40
Do not consume milk	0.10

1. What is the relative risk of bad breath in milk consumers compared to nonconsumers?

2. Suppose the study is a randomized clinical trial (i.e., participants were randomized to drink milk or a placebo). Which of the following inferences would be appropriate based only on these data? Explain your answer. (Assume that the results are statistically significant.)
   a. We cannot conclude that milk consumption is associated with the incidence of bad breath.
   b. There is a real (but maybe not causal) association between milk consumption and the incidence of bad breath.
   c. There may be a causal association between milk consumption and the incidence of bad breath.

3. If this study were a cohort study with a good study design to minimize bias, then from these data alone, which of the following inferences would be appropriate?
   a. We cannot conclude that milk consumption is associated with the incidence of bad breath.
   b. There is a real (but maybe not causal) association between milk consumption and the incidence of bad breath.
   c. There may be a causal association between milk consumption and the incidence of bad breath.

4. If this were a poorly designed cohort study in which there were multiple sources of bias, which of the following would be appropriate?
   a. We cannot conclude that milk consumption is associated with the incidence of bad breath.
   b. There is a real (but maybe not causal) association between milk consumption and the incidence of bad breath.
   c. There may be a causal association between milk consumption and the incidence of bad breath.

Additional results from the study are available in the two tables below:

Incidence of Bad Breath by Milk Consumption (Women Only)

GROUP	INCIDENCE OF BAD BREATH
Consume milk	0.40
Do not consume milk	0.10

Incidence of Bad Breath by Milk Consumption (Men Only)

GROUP	INCIDENCE OF BAD BREATH
Consume milk	0.32
Do not consume milk	0.08

5. For the relationship between milk consumption and incidence of bad breath, gender is:
   a. An effect modifier
   b. A confounder
   c. Both
   d. Neither

Explain your answer.

6. Is it important to adjust for gender when reporting the results of this study? Why or why not?

7. How would you report the results with respect to gender?

14 ASSOCIATION AND CAUSALITY

Epidemiology focuses on determining the causes of disease so that efforts can be made to prevent and control disease and promote and improve health. Often, there are many contributing factors to the development of a disease. This **multifactorial etiology** can make it difficult to determine the exact cause of a disease; thus, epidemiologists must consider an array of evidence and infer causality from this available evidence.

As you now know, epidemiological studies begin with establishing whether there is an **association**, or an identifiable relationship, between an exposure and an outcome. This relationship may be a true relationship between the exposure and the outcome, or a spurious (false) association. A spurious association may occur by chance or might be explained by confounding with a third variable (see Chapter 13 for a full discussion of confounding); therefore, it is important to understand that association does not immediately indicate causation. The finding of a true association does, however, imply that the exposure might cause the disease.

Four types of factors play a role in the causation of disease—predisposing factors, enabling factors, precipitating factors, and reinforcing factors.

- Predisposing factors include characteristics such as sex, age, educational status, marital status, work environment, previous or concurrent illness, and even attitudes toward using health services, or genetic traits that may result in an a weakened immune system or increased susceptibility to a disease-causing agent.
- Enabling factors are the social determinants of health—income, nutrition, housing, and health insurance, among others that can either positively or negatively influence the odds of an individual developing a disease. For example, an individual who works in a job that does not provide health insurance and thus does not receive regular health care and routine screenings may be more at risk of developing a disease than an individual who does have health insurance.
- Precipitating factors are associated with the clear onset of a disease or illness. For instance, in order to develop measles, one must be exposed to the measles virus. In some cases, one must be exposed to a specific amount or level of an infections organism. Some foodborne pathogens, such as *Staphylococcus aureus*, multiply and produce toxins in food. An individual has to ingest an infective dose of the pathogen in order to develop the disease.
- Reinforcing factors include repeated exposures, environmental conditions, or work-related activities or behaviors that aggravate or perpetuate an established disease or an injury. Individuals engaged in repetitive motions of the hands from activities like typing on a computer keyboard for many hours a day may develop carpal tunnel syndrome. A diagnosed individual may not want to or be able to take time off from work to allow the condition to resolve. Continuing to work using the same repetitive motions will worsen the condition.

HENLE-KOCH POSTULATES

The roots of causality began more than 100 years ago. Based on Louis Pasteur's work with microorganisms, Jakob Henle developed and Robert Koch further refined the following postulates:

- The organism is always found with the disease and is not found with any other disease.
- The organism must be able to be isolated and grown in pure culture.
- The organism must, when inoculated into a susceptible animal, cause the specific disease.
- The organism must then be recovered from the animal and identified.

Anthrax, tuberculosis, and tetanus were some of the first diseases to meet these criteria. In fact, many infectious diseases conform to the postulates; however, they are inadequate for chronic disease and even for some communicable diseases. Often, a single factor may be the cause of several diseases. For example, tobacco use—and, in particular, smoking—has been consistently linked to not only lung cancer, but to bladder cancer, cervical cancer, pancreatic cancer, stomach cancer, acute myeloid leukemia, an array of lung diseases including emphysema, and bronchitis, as well as infertility, preterm birth, and low birth weight, among many other diseases. Conversely, some diseases, for example coronary heart disease, are associated with several risk factors, including poor nutrition, alcohol use, lack of exercise, and smoking. It is clear that a more complex method of assessing causal inference is required.

CONSIDERATIONS FOR CAUSATION

Sir Austin Bradford Hill developed a list of criteria that have long been used to provide evidence of causal relations among different risk factors and outcomes (see Chapter 2 for more details about his many accomplishments in epidemiology). Keep in mind that it is very difficult for an association to meet all the criteria; consequently, we must consider all that is known before determining a causal relationship.

TEMPORAL RELATIONSHIP

A temporal relationship is arguably the most important criterion. If a factor is a cause of a disease, the exposure must have occurred before the disease developed. This is easiest to establish in cohort studies. Establishing causality is more problematic in case-control and cross-sectional studies. In case-control studies, exposure data needs to be obtained or re-created from past records, so timing is imprecise. In cross-sectional studies, data on the exposure and the disease are collected at the same time. As a result, establishing temporality is difficult at best. Knowing the **latency period**, or the time between the exposure and the disease, is also important in determining causality. For example, there is a clear link between asbestos and increased risk of lung cancer, but the latency period is at least 15 to 20 years. Thus, if lung cancer develops after only three years of exposure, it is safe to conclude that the lung cancer was not a result of exposure to asbestos.

STRENGTH OF THE ASSOCIATION

The stronger the association, as measured by the relative risk or odds ratio, the more likely it is that the relationship is causal. Typically, in epidemiology, a relative risk greater than two can be considered strong. A weak association, however, does not preclude a causal relationship. Sometimes weak associations occur when there are a number of other possible causes. For example, there are numerous contributors to coronary heart disease, so any one factor may not have an exceptionally strong association with coronary heart disease. Since we know there are multiple causes of coronary heart disease, we would not exclude a factor as a cause just on the basis of the strength of the association.

DOSE-RESPONSE RELATIONSHIP

As the dose of the exposure increases, the risk of the disease increases. For instance, as the number of cigarettes a person smokes a day increases, so does the risk of developing lung cancer. If a dose-response relationship is present, it is evidence for causation. But if this does not exist, it does not necessarily rule out a causal relationship. There may be a threshold, so disease may not develop if this threshold of exposure is not reached. Above the level, disease may develop. For example, at certain levels, radiation is therapeutic, but over a certain level, radiation sickness can develop.

REVERSIBILITY

If a factor causes a disease, we expect the risk of the disease to decline when exposure is reduced or eliminated. For example, quitting smoking is associated with a reduction in lung cancer risk relative to continuing to smoke. This finding indicates that the exposure is most likely causal. It is important to note that in some cases, pathogenic processes may be irreversibly started; as a result, stopping the exposure will not prevent the disease from occurring.

CONSISTENCY

If an association is causal, we expect to see it consistently, across different studies and in different appropriate populations. Replication of the findings is very important and should be evident across a wide array of study designs. This also points to the importance of periodically pooling the results of these studies through meta-analysis or systematic review. **Meta-analysis** combines the results of a number of studies using statistical methods to pool the results, providing a quantitative summary. A **systematic review** uses a variety of rigorous strategies to find, analyze, and synthesize all relevant studies on a particular topic.

PLAUSIBILITY

As association that is consistent with the current body of biological knowledge is likely to be causal. Essentially, we are determining whether the exposure/outcome relationship is logical and reasonable, given what we know. Consistency with biological knowledge makes epidemiological findings easier to interpret.

Sometimes we have epidemiological findings that precede biological knowledge

supporting the relationship, so this criterion supports a causal relationship, but does not preclude one. For instance, it is now known that some viruses contracted by a pregnant woman can cross the placenta and affect the unborn baby's growth and development. These are called teratogenic viruses and one of the most devastating is rubella. In adults, rubella is a mild illness, characterized by a red rash and fever that is transmitted through coughing and sneezing. Babies born to mothers who contracted rubella early in their pregnancy are often born with severe congenital malformations affecting vision, hearing, the heart and the brain.

In 1941, well before teratogenic viruses were discovered, Sir Norman Gregg, an ophthalmologist in Australia, noticed that mothers who contracted rubella during pregnancy gave birth to babies who had severe congenital cataracts. The mechanism by which rubella affected the unborn babies' eyesight was not specifically known, and the actual virus itself was not isolated until 1961. Clearly we would not ignore this association simply because we do not why it happened. This demonstrates that biologic plausibility is useful for determining causality but not absolutely necessary. If we don't see results consistent with what we would expect, we are likely to be more demanding regarding the requirements about size and significance of the differences observed and have the study replicated in other populations and by other researchers.

COHERENCE

The association should be compatible with existing theory and consistent with other knowledge. The theory and knowledge can be from any applicable domain: biological, social, epidemiological, or statistical. If a relationship is causal, we expect the findings to be consistent with other data from other important areas. For instance, we would expect to see a decrease in oral cancer cases if a decrease in chewing tobacco sales is reported. If fewer people are chewing tobacco, the number of people diagnosed with oral cancer would decrease, providing evidence of a causal relationship.

CONSIDERATION OF ALTERNATE EXPLANATIONS

A very important step in determining a causal relationship is considering whether there are any other possible explanations. Investigators should determine whether any possible confounders have been ruled out. Recall that a confounder is a variable that is also related to the exposure or outcome that may affect results. For example, if you are looking at the relationship between asbestos and lung cancer, smoking status may muddy up the relationship. We must also consider to what extent other potential explanations have been explored, acknowledged, or statistically controlled.

SPECIFICITY OF THE ASSOCIATION

Harkening back to the Henle-Koch postulates, an association is specific when a certain exposure is associated with only one disease. This is perhaps the weakest criterion because, as previously noted, there are many exposures that are causally related to an array of outcomes. When this criterion is met, it lends additional support, but as with dose-response relationship, absence of this does not negate a causal relationship.

STUDY DESIGN

Different types of studies provide differing levels of evidence regarding causality. Well-designed randomized controlled trials provide the best evidence of causality, but usually we use these studies when looking at prevention and treatment as ethical considerations often preclude researchers from using this type of design. It would not be ethical to ask one group to begin smoking and then ask another group to refrain to determine whether cigarette smoking causes cancer. Consequently, most causal evidence comes from observational studies.

Since, in most cases, we cannot ethically use randomized controlled trials to study causality, the next best evidence comes from cohort studies. Bias in these studies tends to be minimized because participants are free of the disease at the beginning, and the researchers are generally able to confirm exposure as well.

In the event that cohort studies are not available, likely due to their time and financial constraints, case-control studies are useful in providing evidence of causality. Case-control studies are more subject to bias, particularly recall and selection bias, than are cohort studies. If the study is well designed, a case-control study can provide solid evidence of causal associations. Often, case-controls studies will provide the basis for a follow-up cohort study, as occurred in Richard Doll and Austin Bradford Hill's series of studies that determined the causal relation between smoking and lung cancer (see Chapter 2 for more about the work of Doll and Hill).

Working in the United Kingdom, Doll and Hill conducted a case-control study in 1947, comparing the smoking habits of lung cancer patients with those patients who did not have lung cancer. The results of this study indicated that the odds of being a smoker were nine times higher in those with lung cancer, as compared to those without lung cancer. A cohort study was conducted in 1951 with information obtained from the list of doctors listed in the British Medical Registry. Information about present and past smoking habits was gathered. Lung cancer mortality data came from death certificates issued in the following years. The mortality rate from lung cancer was 18.4 times higher in smokers as compared to nonsmokers.

Cross-sectional studies are less useful in determining causality, as they do not provide direct evidence of temporality. Cross-sectional studies are a snapshot in time, with the data on the exposure and the outcome collected at the same time. Depending on how a questionnaire is developed, we may be able to infer some time sequence, but the evidence will be weak.

The weakest evidence of causality comes from ecological studies. Since data is collected at the aggregate or group and not at the individual person level, it is incorrect to assume causal relations between an exposure and an outcome in an ecological study. Making this assumption is called an **ecological fallacy**. Ecological studies do help quantify the effects of phenomena that are difficult to measure at the person level—air pollution, pesticide residue, and fluoride in drinking water, for example.

15 SURVEILLANCE, PREVENTION, AND SCREENING

The Centers for Disease Control and Prevention (CDC) define **public health surveillance** as the:

> "ongoing systematic collection, analysis, and interpretation of health data essential to planning, implementation, and evaluation of public health practice, as well as the timely dissemination of these data to those who need to know."

Three types of surveillance comprise the U.S. and global public health surveillance system:

- Passive surveillance;
- Active surveillance;
- Syndromic surveillance.

Passive surveillance requires that medical care providers—not public health practitioners—report notifiable diseases on a case-by-case basis to state and local health agencies. Reportable diseases are usually infectious and have the potential to endanger a population. Examples include most sexually transmitted infections, including HIV; diseases prevented by vaccines such as measles, polio, and pertussis, among others; and illnesses associated with outbreaks such as food-borne diseases like *E. coli* and *salmonella*. In certain regions, diseases of local concern can be listed as reportable. For example, in the northeast and mid-Atlantic regions of the United States, Lyme disease, which is transmitted by the deer tick, is reportable. In areas of the southwest, Lyme disease is less of a concern and is therefore not reportable.

Often, reportable diseases captured through passive surveillance are reported to disease registries. These registries, such as the Surveillance, Epidemiology, and End Results (SEER) program at the National Cancer Institute, are centralized databases that are used to collect data about a particular disease. Disease registries are used to collect incidence, prevalence, and survival data. With this information, we are able to:

- Monitor trends over time
- Determine disease patterns in populations
- Plan and evaluate programs to reduce disease surveillance
- Prioritize health needs
- Conduct clinical, epidemiological, and health-services research

Passive surveillance results in an accumulation of case reports, from which trends in time and rates in specific populations can be monitored. Because passive surveillance relies on medical care providers reporting disease incidence, our data will always be incomplete. Additionally, passive surveillance will only include data from individuals who seek medical care.

Active surveillance, on the other hand, is initiated and completed by public health agency staff. These public health practitioners contact healthcare providers, laboratories, and others to identify possible cases. This type of surveillance is both time consuming and labor intensive.

Federal and state health agencies engage in a host of active surveillance projects to protect the public's health through education, policy, and programs. An example of an ongoing active surveillance project that is conducted at the state level is the Behavioral Risk Factor Surveillance Survey, better known as the BRFSS (www.cdc.gov/brfss). Via telephone survey of a random selection of individuals, the BRFSS collects behavioral risk factors for chronic diseases and injuries. The survey collects information on behaviors such as alcohol consumption, smoking habits, preventive health practices such as diet, exercise, and seatbelt use, as well as health care access.

Active surveillance is often used during an outbreak, for instance to investigate a possible childhood cancer cluster in a small town, or a high-profile event. Active surveillance is also used when there is a need to find all the cases of a specific disease, often for research.

The final type of surveillance used is **syndromic surveillance**. This type uses symptom information to alert public health officials to a potential problem. For instance, data from pharmaceutical sales may be used to determine if there is a rise in the purchase of antidiarrheal drugs. This rise might be the first sign of a food-borne outbreak of disease. The public health community uses syndromic surveillance as one of the primary indicators of a bioterrorism attack. This type of surveillance can provide timely (pre-diagnosis) data; however, getting that data and using it to predict public health outcomes is both labor- and time-intensive.

DISEASE PREVENTION

In addition to collection data and monitoring disease trends, public health practitioners want to prevent disease from occurring in the first place. The public health community breaks **prevention** into three levels:\

- Primary prevention;
- Secondary prevention;
- Tertiary prevention.

Primary prevention aims to prevent the occurrence of disease through personal and community efforts such as health education, improved nutrition, immunizations, sanitation, and infection control. **Secondary prevention** attempts to reduce the progress of disease through early detection and prompt interventions. Often, this occurs through screening tests. For example, mammography is used to diagnose breast cancer before clinical symptoms are present in order to improve prognosis and prevent the cancer from progressing and spreading. Finally, **tertiary prevention** focuses on reducing impairment and minimizing the suffering of patients with disease through rehabilitation and palliative care. Tertiary services are provided to patients who already have the disease of interest. Rehabilitation or treatment is used to restore this individual to a useful, productive, and satisfying lifestyle, given the extent of the disease.

THE ROLE OF EPIDEMIOLOGY IN SECONDARY PREVENTION

One of the most common forms of secondary prevention of disease is screening. Screening is the presumptive identification of an unrecognized disease by the application of tests or other procedures that can be applied rapidly. It focuses on early detection of a disease, which can lead to more favorable outcomes. Examples of screening tests commonly used are the PPD test to detect tuberculosis; the Pap smear to detect cervical cancer; and the PSA test to detect prostate cancer, among others.

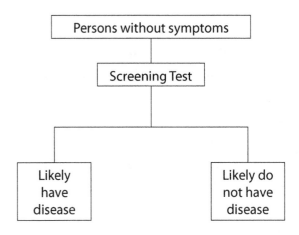

Figure 15.1   Illustration of Screening Test

A set of five criteria exists in order to conduct a screening test:

- The test must address an important health problem;
- There is a known natural history of the screened for disease;
- A recognized latent stage of disease must exist;
- An acceptable form of treatment must exist;
- Early treatment must improve the outcome.

In order to implement a screening program into a population at large, the test must meet the criteria listed above, and there must be:

- A reasonable cost for the test;
- Diagnostic/treatment facilities available to those being screened;
- Policies, procedures and threshold levels on tests should be determined in advance to establish who should be referred for further testing, diagnostics and possible treatment;
- The process should be simple enough to encourage people to participate.

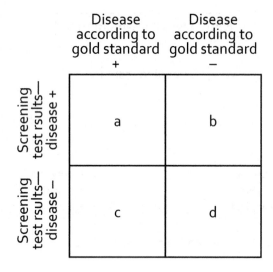

Figure 15.2   2×2 Table for Screening Test

From an epidemiological perspective, a screening test can be viewed through a 2×2 table structure.

The gold standard here is a measure that tells us what is "true" about the individual's health status. Often, this gold standard is the accepted method for diagnosing the disease.

We interpret the cells of this 2×2 table as follows:

A = true positive screen
B = false positive screen
C = false negative screen
D = true negative screen

	PSA ≤7 ng/mL		
	Prostate Cancer	No Cancer	Total
PSA+	(TP)175	(FP)50	225
PSA−	(FN)80	(TN)415	495
Total	255	465	720

Figure 15. 3   Example Screening 2×2 table

Using these cells, we can estimate the sensitivity and specificity of the screening test. **Sensitivity** is the ability of a screening test to correctly identify individuals with the disease. It is calculated as the proportion of individuals with the disease who have a positive test: a/(a+c). Once this calculation has been made, we multiply it by 100 to get a percent.

Specificity is the ability of a screening test to correctly identify individuals without the disease. It is calculated as the proportion of individuals without disease who have a negative test: d/(b+d). Again, we multiply our finding here by 100 to get a percent. Sensitivity and specificity tell us how well the screening test does at correctly identifying individuals with and without the disease in a population. This numbers are often multiplied by 100 to create percentages, which are more easily understood by the general population.

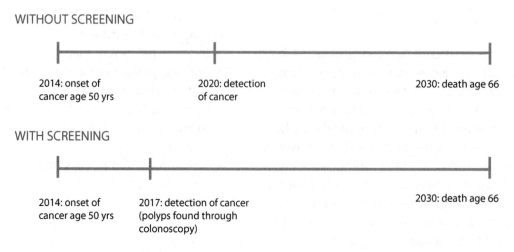

Figure 15.4   Illustration of Lead Time Bias

For instance, supposed we have the following, using a PSA test to screen for prostate cancer:

We calculate as follows:

Sensitivity (TP/TP+FN)	175/255 = 69% Of those who truly had prostate cancer, 69% tested positive
Specificity (TN/TN+FP)	415/465 = 89% Of those who truly did not have prostate cancer, 89% tested negative

If we want to tell an individual who has completed a screening test about the probability of having or not having the disease given a positive or negative test, we want to report the **positive or negative predictive value**. The positive predictive value tells us the proportion of individuals with a positive screen who truly have the disease. The negative predictive value is the proportion of individuals with a negative screening test who truly do not have the disease. Using our 2x2 table, the positive predictive value (PPV) and negative predictive value (NPV) are calculated by:

PPV (TP/TP+FP)	175/225 = 78% Of those who **TESTED** positive, 78% actually have prostate cancer
NPV (TN/TN+FN)	415/495 = 84% Of those who **TESTED** negative, 84% actually do not have prostate cancer

The PPV and NPV can also help us determine when to begin screening for maximum public health impact. For instance, because the prevalence of prostate cancer is higher in older men, we are more likely to catch more true positives using a PSA screening test in that group than in younger populations.

Ideally a screening test will have high sensitivity, high specificity, high PPV and high NPV; however, this is almost impossible to achieve. If a screening test is more sensitive, it is less specific and conversely, the more specific it is, the less sensitive it is. Imagine a screening test where those who score above a specific score (above the line in the shaded areas in Figure 15.5 below) are said to have a positive result. In each of the three figures below, there will be:

True positives → the + in the shaded area above the line

True negatives → the − in the area below the line

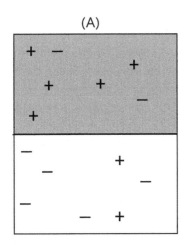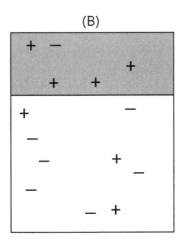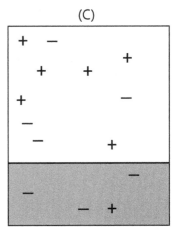

Figure 15.5 Illustration of relationship between sensitivity, specificity, false positives, and false negatives in a screening test

False positives → the − in the shaded area above the line

False negatives → the + in the area below the line

If we were to raise the cut-off point of the screening test (B in Figure 15.5), we will increase the specificity (percent of negatives classified as negative) and decrease the sensitivity. There will be fewer false positives, but some people who have the disease will screen negative and believe they are healthy. If we lowered the cut-off point (C in Figure 15.5), we will increase the sensitivity (percent of positives classified as positive) and decrease the specificity. There will be fewer false negatives, but some people who are healthy will screen positive and think they have disease.

Continuing with our prostate cancer example, the potential benefits of prostate cancer screening are obvious as screening can detect cancer early and treatment for prostate cancer may be more effective when it is found early. There are also potential risks of prostate cancer screening, including false positive test results that lead to further tests which may be invasive, painful and expensive and may cause the individual additional anxiety. Screening may also result in the treatment of some prostate cancers that may have never affected a man's health even if left untreated. Prostate cancer treatment may also lead to serious side effects such as impotence (inability to keep an erection) and incontinence (inability to control the flow of urine, resulting in leakage).

BIASES ASSOCIATED WITH SCREENING TESTS

There are three types of bias commonly associated with screening tests:

- Lead-time bias;
- Length bias;
- Volunteer bias.

Lead-time bias occurs when screening detects a disease earlier in its course than if screening had not been performed. The length of time from diagnosis to death is increased. However, the length of life may not increase. Imagine a colon cancer screening test such as colonoscopy. In this example, two 50-year-old women have the same date of onset of cancer.

The two women in this example live to be the same age, and both develop colon cancer. The woman who had the colonoscopy lived three more years with the knowledge of her diagnosis, but this did not increase her survival. Since the screening test in this example had no impact on the outcome (death), and the main goal of screening is to detect disease early so that it is more successfully treated, this suggests that this screening test should not be continued.

To minimize the effects of lead-time bias, the estimates of survival following diagnosis must be corrected for the average lead time.

Length bias is the overestimation of survival duration among screening-detected cases, which is caused by the (relative) excess of slowly progressing cases. These cases—with slowly progressing diseases—are disproportionately identified through screening because the probability of detection is directly proportional to the length of time during which the disease is detectable. It is, therefore, inversely proportional to the rate of progression. In short, people with less aggressive diseases: 1) survive longer; 2) are overrepresented among cases identified by screening; and 3) have better survival after screening. For example, if a 60 year old women finds an early-stage, small tumor in her breast through a routine mammogram, she is not likely to die from the disease. She will be treated before the cancer spreads.

Volunteer bias occurs because people who are healthier, health-conscious, or have medical insurance are more likely to be screened than those who are less healthy or lack insurance. Screening test volunteers are not a random sample from the larger population. They are systematically different from the population as a whole—representing those who have insurance and are more health-conscious. Volunteer bias is a type of selection bias.

EXERCISES

For each of the situations below, draw a 2x2 table and answer the questions.

1. A hypothetical test is developed to screen for asthma. A total of 200 individuals undergo the test. According to the gold standard, 75 have asthma and 125 do not. The test is positive for 55 of those individuals with asthma; 25 of those without asthma tested positive.

    Given the information above, can you complete a 2x2 table? If so, what are the sensitivity, specificity, and positive and negative predictive values of the test? If you cannot complete a table, what information is given, and what is missing?

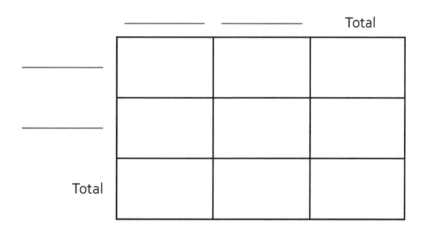

2. Another hypothetical test is developed to screen for influenza. A total of 1000 individuals undergo the screening test. Seven hundred individuals test positive for the screening test; 300 test negative. Of those, 500 truly have influenza and 500 do not.

Given the information above, can you complete a 2×2 table? If so, what are the sensitivity, specificity, and positive and negative predictive values of the test? If not, what information is given, and what is missing?

	_____	_____	Total
_____			
_____			
Total			

16

MATHEMATICAL CONCEPTS

FRACTIONS, DECIMALS, AND PERCENTS

Fractions, decimals, and percents are all numbers that represent a part of the whole. In epidemiology, we often need to convert from fractions to decimals or percents or from decimals to percents or fractions.

Example:

$$\frac{1}{4} = .25 = 25\%$$

Converting a Fraction to a Decimal

Fractions are composed of numerators (the top number) and denominators (the bottom number). To convert a fraction to a decimal, divide the numerator by the denominator.

Example:

$$\frac{3}{8} = 3 \div 8 = .375$$

Converting a Decimal to a Percent

Multiply the decimal by 100, and add a percent sign. We are moving the decimal point two places to the right.

Example:

$$.375 \times 100 = 37.5\%$$

Converting a Percent to a Decimal

Remove the percent sign and divide the percentage by 100. We are moving the decimal point two places to the left.

Example:

$$37.5\% = \frac{37.5}{100} = .375$$

Converting a Percent to a Fraction

Remove the percent sign, put the number over 100 and reduce (i.e., simplify the fraction so that the top number and bottom number cannot be divided by the same number).

Example:

$$56\% = \frac{56}{100} = \frac{14}{25}$$

Convert a Decimal to a Fraction

Determine how many numbers follow the decimal place (i.e., the number of decimal places). If there is one decimal place, place the number over 10 and reduce. If there are two decimal places, place the number over 100 and reduce. If there are three decimal places, place the number over 1000 and reduce.

Example:

One decimal place: $.4 = \frac{4}{10} = \frac{2}{5}$

Two decimal places: $.08 = \frac{8}{100} = \frac{2}{25}$

Three decimal places: $.965 = \frac{965}{1000} = \frac{193}{200}$

MULTIPLYING AND DIVIDING FRACTIONS

To multiply fractions, simply multiply across the numerators of both fractions (the top numbers) and across the denominators of the fractions (the bottom numbers).

Example:

$$\frac{1}{3} \times \frac{5}{8} = \frac{1 \times 5}{3 \times 8} = \frac{5}{24}$$

To divide fractions, take the reciprocal of the second fraction (i.e., "flip" the second fraction) and multiply it by the first. This is also called cross-multiplying.

Example:

$$\frac{1}{3} \div \frac{5}{7} = \frac{1}{3} \times \frac{7}{5} = \frac{7}{15}$$

PERCENT

Find One Number as a Percent of Another:

There are two methods we can use to accomplish this.

Example: The number 15 is what percent of 48?

1. Divide the first number by the second number. Move the decimal two places to the right and add a percent sign. This is the same as multiplying by 100 to convert the decimal to a fraction.
   Divide 15 by 48

$$15 \div 48 = .3125 \ or \ \frac{15}{48} = .3125$$

Move the decimal two places to the right and add a percent sign: 31.25%

Thus, 15 is 31.25% of 48.

2. We can also set this up as a ratio.

$$\frac{15}{48} = \frac{x}{100}$$

Cross multiplying, we get

$$48\,x = 1500$$

Now, solve for x by dividing both sides of the equation by 48

$$x = \frac{1500}{48}$$
$$x = .3125$$

Move the decimal two places to the right or multiply by 100 to convert the decimals into a percent. Do not forget to add the percent sign.

$$.3124 \times 100 = 31.25\%$$

Find a Percent of a Given Number

Convert the given percent to a decimal. Multiply the number by that decimal.

Example: What is 67% of 205?

Convert 67% into a decimal by moving the decimal two places to the left or by dividing by 100:

$$67\% = .67$$

$$\frac{67}{100} = .67$$

Multiply 205 by 0.67: 205 * 0.67 = 137.35

Thus 137.35 is 67% of 205.

RATIOS, PROPORTIONS, AND RATES

In introductory epidemiology, we will most often compare two-category variables (dichotomous variables) using ratios, proportions, and rates. All three measures are based on the same formula

ratio, proportion, rate = x/y

In this formula, x and y represent two quantities we are comparing. It simply means x is divided by y.

Some key points when working with epidemiological data:

- Denominators are reflective of population size; everyone must have potential to enter the numerator group.
- Numerators are reflective of those in the population who are exposed or unexposed, or diseased or not diseased.

- A **rate** is a measure of the frequency with which an event occurs in a defined population in a defined time.
    - o The number of cases of measles in the United States in 2012
- A **ratio** is used to compare the occurrence of a variable in two different groups. These groups can be completely independent of each other.

$$\frac{Males\ receiving\ polio\ immunization}{Females\ receiving\ polio\ immunization}$$

- A **proportion** is a type of ratio in which the numerator is a subset of the denominator.

$$\frac{Males\ receiving\ polio\ immunization}{All\ individuals\ immunized\ against\ polio}$$

More About Ratios

For example, suppose 250 individuals received an HPV immunization; 175 were females, 75 were males. We could look at the ratio as two completely independent groups, comparing the number of females immunized to the number of males immunized. These ratios are often reduced by dividing, so that one value is 1.

$$\frac{females}{males} = \frac{175}{75} = 2.3:1$$

This tells us that just over two females are vaccinated for every one male vaccinated.

We can also look at females as a subset of the entire group immunized. This is a type of ratio called a proportion. These are often expressed as a percentage.

$$\frac{Females}{All} = \frac{175}{250} = .70 \quad 100 = 70\%$$

Thus, 70% of the individuals immunized were females.

Ratios—Independent x and y

During the first year of a national surveillance study of Lyme disease, the CDC received 2365 case reports. There were 1572 cases in females; 793 cases were male. Here is how to calculate the female to male ratio:

1. Define x and y	x = cases in females y = cases in males
2. Identify x and y	$x = 1572$ $y = 793$
3. Set up the ratio x/y	1572/793
4. Reduce the fraction, so that one value equals 1	Females/males 1572/793 = 1.98/1
5. Express the ratio in any of the these ways	1.98 to 1 1.98:1 1.98/1
6. Explain the relationship in words	There are almost two female Lyme disease cases for each male Lyme disease patient reported to the CDC.

Proportions—x Included in y

Based on the same data, we can calculate the proportion of Lyme disease cases that were males.

1. Define x and y	x = cases in males y = all cases
2. Identify x and y	x = 793 y = 2365
3. Set up the ratio x/y	793/2365
4. Reduce the fraction by dividing the smaller number by the larger number	Males/all cases 793/2,365 = .34
5. Proportions are usually expressed as percentages, so multiply by 100	.34 × 100 = 34%
6. Explain the relationship in words	About 34% of the reported Lyme disease cases were in males.

ROUNDING

Precision is very important in epidemiology. It is always best to calculate all measures to the third decimal point (the thousandths place) and round back to two decimals (the hundredths place) for the greatest accuracy. When rounding with decimals, the following rules apply:

- Determine what your rounding digit is, and look to the right side of it. If that digit is 4, 3, 2, or 1, simply drop all digits to the right of it. This is rounding down.
- Determine what your rounding digit is, and look to the right side of it. If that digit is 5, 6, 7, 8, or 9, add one to the rounding digit and drop all digits to the right of it. This is rounding up.

Example:

Rounding down to the nearest hundredth: 123.952 becomes 123.95

Rounding up to the nearest hundredth: 65.748 becomes 65.75

17 BIOSTATISTICAL CONCEPTS

Biostatistics is a part of statistical theory and methods, which are directly related to investigations in the biological and health sciences. Statistics is the art and science of collecting, organizing, describing, and analyzing data.

Statistical methods provide principled and objective methods for:

1. Testing scientific hypotheses;
2. Weighing evidence;
3. Estimating risk and other characteristics of a population.

In most epidemiological studies (both descriptive and analytic), we study a sample—a subset of a larger population—and use both descriptive and inferential statistics to make generalizations about the larger population (known as our source population).

Descriptive statistics describe important features and trends in a data sample and allow us to decide whether or not it is representative of the source population. **Inferential statistics** are used to investigate the research hypothesis about the source population, using information from the sample.

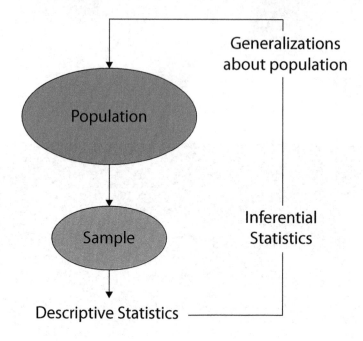

Figure 17.1   Illustration of the Application of Statistics

TYPES OF VARIABLES

Each epidemiological study will collect data about the individuals participating in the study. Each characteristic of each individual is called a variable.

Variables can be either categorical or quantitative. Categorical variables place individuals into one of several groups or categories with no numerical values. Categorical variables can be **nominal** (named categories), **ordinal** (ranked categories), or **binary** (dichotomous with only two possible levels). **Quantitative variables**, on the other hand, take on a numerical value. They can be either **continuous** or **discrete** (count).

We can categorize our study variables into independent and dependent variables. The **independent variable**, also known as the explanatory variable, is the treatment, exposure, or risk factor(s) being studied. The **dependent variable**, commonly known as the response variable, is the health outcome or response being investigated.

After the appropriate measure of association is calculated, we can use statistical analyses to first determine if we have evidence to reject or accept the **null hypothesis**, which states that there is no relationship between our variables of interest. This type of statistical analysis is known as **inferential statistics**. By definition, inferential statistics allow us to conduct an investigation of a **research hypothesis**, which states the relationship we expect to find between our variables in a population of interest, using information from a random sample of data from that population.

DETERMINING STATISTICAL SIGNIFICANCE

Using inferential statistical techniques, we are able to quantify the strength of association between a disease and risk factor in two ways:

- *p*-value;
- Confidence intervals.

A ***p*-value** is defined as the probability of obtaining a result at least as extreme as that observed in the study by chance alone. In other words, a *p*-value is a measure of compatibility between a hypothesis and the data. A *p*-value is computed from, and will vary with, the data that is collected. The conventionally accepted p-value for health sciences research, in most cases, is 0.05, indicating that the probability that the result found in our data set occurred by chance is less than 5%. When we are considering more serious outcomes such as debilitating side effects or death, we often set the *p* value at .01, reducing the probability of a result occurring by chance to 1%.

At the design phase of an epidemiological study, the research team will define the α-level (alpha level) of the study. This α-level is defined as the probability of observing an association between the disease and exposure of interest, when the truth is that association does not exist. In short, the α-level determines how likely it will be that we find an association between the disease and exposure in our study sample, when, in fact, that association does not exist in the larger population.

At the end of our study analysis, we compare our calculated *p*-value to the **α-level** you set at the beginning of the study.

If our p-value is *less than* our α-level, we say the finding is statistically significantly different, indicating that there is a likely association between the exposure and the disease.

If our p-value is *greater than* our α-level, we say the finding is not statistically significant, or that there is no association between the disease and exposure.

The second way we quantify the strength of association between a measure of association with disease and exposure is by constructing confidence intervals. **Confidence intervals** (CI) measure the range within which the true

magnitude of effect lies with a certain degree of certainty.

Like studies that calculate *p*-values, studies that calculate confidence intervals require that the α-level be set at the beginning of the study. The α-level determines the range of the confidence interval. For example, when α is set at 0.05, we construct a 95% confidence interval. If α is 0.10, the corresponding confidence interval is 90%.

Interpreting a confidence interval can be tricky. For a 95% confidence interval, we are 95% confident that across repeated samples, the population measure of association between disease and exposure is contained in the interval. The correct ways to interpret a confidence interval for a relative risk that ranges from 2.5–5 are:

- Based on a 95% confidence interval, this data is consistent with a population relative risk between 2.5 and 5.

- We've used a procedure under unbiased random sampling procedures to calculate an interval that contains a relative risk in 95% of samples. For this particular sample, that interval is 2.5 to 5.

If the confidence interval contains the null value, where the measure of association shows no relationship between disease and exposure, we conclude that the relationship between the disease and exposure is not statistically significant. For instance, we may obtain a relative risk of 1.5 with a confidence interval ranging from .95 to 2.6. The null or no association value for a relative risk is 1.0. Our confidence interval contains 1.0, thus we conclude that there is no association between our exposure and our outcome. It also helps to consider that it appears that the relationship is sometimes protective (.95) and sometimes risky (2.6) within the same group, which does not seem logical. Thus it is likely that there is no relationship between the two variables of interest.

If, on the other hand, the confidence interval does not include the null value, we say the relationship is statistically significant. These statistically significant associations can be either hazardous or protective.

SAMPLE SIZE

It is important to note that the *p*-value and the confidence interval are greatly affected by the sample size. A larger sample size results in a narrower confidence interval, which provides additional assurance that our measure of association is a good estimate of the population. When it comes to sample size, bigger is generally better; however caution should be used when assessing relationships in very large studies. In very large studies, almost all associations will be significant. This is why it is imperative that researchers hypothesize expected relationship based on previous research prior to conducting statistical analyses. This prevents capitalizing on chance associations that may occur in large datasets.

REFERENCES

Badamchi, A. & Walters, A. (2014). Vaccine developments. *British Society for Immunology.* Retrieved from http://bitesized.immunology.org/vaccines-trerapeutics/vaccine-developments.

Baxter, N. N., Goldwasser, M. A., Paszat, L. F., Saskin, R., Urbach, D. R., & Rabeneck, L. (2009). Association of colonoscopy and death from colorectal cancer. *Annals of Internal Medicine, 150*(1), 1–8.

Best, M., & Neuhauser, D. (2004). Ignaz Semmelweis and the birth of infection control. *Quality & Safety in Health Care, 13,* 233–234.

Bollet, A. J. (2004). *Plagues & Poxes: The Impact of Human History on Epidemic Disease.* New York: Demos Medical Publishing.

Bonita, R., Beaglehole, R., & Kjellstrom, T. (2006). *Basic Epidemiology.* 2nd Edition. Geneva: World Health Organization.

Brooks, J. (1996). The sad and tragic life of Typhoid Mary. *Canadian Medical Association Journal, 154* (6), 915–916.

Brown, J. R., & Thornton, J. L. (1957). Percivall Pott (1714–1788) and Chimney Sweepers' cancer of the scrotum. *British Journal of Industrial Medicine, 14* (1), 68–70.

Brown, S. R. (2003). *Scurvy: How a Surgeon, a Mariner, and a Gentleman Solved the Greatest Medical Mystery of the Age of Sail.* New York: St. Martin's Press.

Centers for Disease Control and Prevention (2014). Colorectal (colon) cancer: What should I know about screening? Retrieved from http://www.cdc.gov/cancer/colorectal/basic_info/screening/

Centers for Disease Control and Prevention (2014). Behavioral risk factor surveillance system. Retrieved from http://www.cdc.gov/brfss/data_documentation/index.htm

Centers for Disease Control and Prevention. (2003). Cigarette smoking and lung cancer. Case Studies in Applied Epidemiology No. 731703.

Doll, R. (1994). Austin Bradford Hill. *Biographical Memoirs of Fellows of the Royal Society, 40,* 128–140.

Dunn, P.M. (2007). Perinatal lessons from the past: Sir Norman Gregg, CHM, MC, of Sydney (1892–1966) and rubella embryopathy. *Archives of Disease in Childhood Fetal Neonatal Edition, 91,* F513–F514.

Emanuel, E.J., Grady, C. Crouch, R.A., Lie, R.K, Miller, F.G. and Wendler, D. (2008). *The Oxford Textbook of Clinical Research Ethics.* New York: Oxford University Press.

Encina, G.B. (2004). Milgram's experiment on obedience to authority. Retrieved from http://www.cnr.berkeley.edu/ucce50/ag-labor/7article/article35.htm

Evans, A. A. (1976). Causation and disease: The Henle-Koch postulates revisited. *Yale Journal of Biology and Medicine, 49,* 175–195.

Etzioni, R., Tsodikov, A., Mariotto, A., Szabo, A., Falcon, S., Wegelin, J., ... & Feuer, E. (2008). Quantifying the role of PSA screening in the US prostate cancer mortality decline. *Cancer Causes & Control, 19*(2), 175–181.

Friis, R.H. (2010). *Epidemiology 101.* Boston: Jones & Bartlett Publishers.

Frerichs, R. R. John Snow. Retrieved December 7, 2013, from http://www.ph.ucla.edu/epi/snow.html

Gerstman, B. B. (1998). *Epidemiology Kept Simple.* New York: Wiley-Liss.

Gigerenzer, G., Mata, J., & Frank, R. (2009). Public knowledge of benefits of breast and prostate cancer screening in Europe. *Journal of the National Cancer Institute, 101*(17), 1216–1220.

Gorell, J.M., Johnson, C.C., Rybicki, E.L., Peterson, E.L., Richardson, R.J. (1998). The risk of Parkinson's disease with exposure to pesticides, farming, well water, and rural living. *Neurology, 50(5),* 1346–1350.

Gordis, L. (2009). *Epidemiology.* 4th Edition, Philadelphia: Saunders Elsevier.

Gross, C. P., McAvay, G. J., Krumholz, H. M., Paltiel, A. D., Bhasin, D., & Tinetti, M. E. (2006). The effect of age and chronic illness on life expectancy after a diagnosis of colorectal cancer: Implications for screening. *Annals of internal medicine, 145*(9), 646–653.

Hayat, M. J., Howlader, N., Reichman, M. E., & Edwards, B. K. (2007). Cancer statistics, trends, and multiple primary cancer analyses from the Surveillance, Epidemiology, and End Results (SEER) Program. *The Oncologist, 12*(1), 20–37.

Hempel, S. (2007). *The Strange Case of the Broad Street Pump: John Snow and the Mystery of Cholera.* Berkeley, CA: University of California Press.

Hoyert, D.L, & Xu, J. (2012). Deaths: Preliminary data for 2011. *National Vital Statistics Reports, 61* (6).

Jacobsen, K.H. (2008). *Introduction to Global Health.* Sudbury, MA: Jones and Bartlett.

Jean-Franiois Viel, Bruno Challier. (1995). Bladder cancer among French farmers: Does exposure to pesticides in vineyards play a part? *Occupational and Environmental Medicine* 1995;52:587–592

Johnson, N.P.A.S. & Mueller, J. (2002). Updating the accounts: Global mortality of the 1918–1920 "Spanish" influenza pandemic. *Bulletin of the History of Medicine, 76* (1), 105–115.

Kipling, M. D., & Waldron, H. A. (1975). Percivall Pott and cancer scrota. *British Journal of Industrial Medicine, 32,* 244–250.

Marmot, M. G., Smith, G. D., Stansfeld, S., Patel, C., North, F., Head, J., Brunner, E., & Feeney, A. (1991). Health inequalities among British civil servants: The Whitehall II studies. *Lancet,* 337 (8754), 1387–1393.

Merrill, R.M. (2011). *Principles of Epidemiology Workbook.* Sudbury, MA: Jones and Bartlett.

Merrill, R.M. (2010). *Introduction to Epidemiology.* 5th Edition. Boston: Jones and Bartlett.

National Oceanic and Atmospheric Administration. (2011). Great Tohoku Japan earthquake and tsunami, 11 March 2011. Retrieved from http://www.ngdc.noaa.gov/hazard/honshu_11mar2011.shtml

Nugent, R. (2008). Chronic diseases in developing countries: Health and economic burdens. *Annals of the New York Academy of Science, 1136,* 70–79.

Nyström, L., Wall, S., Rutqvist, L. E., Lindgren, A., Lindqvist, M., Ryden, S., ... & Larsson, L. G. (1993). Breast cancer screening with mammography: Overview of Swedish randomised trials. *The Lancet, 341*(8851), 973–978.

Patterson, K. B., & Runge, T. (2002). Smallpox and the Native American. *American Journal of Medical Science, 323* (4) 216–222.

Peto, R., & Beral, V. (2010). Sir Richard Doll CH OBE. *Biographical Memoirs of Fellows of the Royal Society, 56,* 63–83.

Porta, M. (2008). *A dictionary of epidemiology.* Oxford, New York: Oxford University Press,

Rothman, N., Smith, M. T., Hayes, R. B., Traver, R. D., Hoener, B. A., Campleman, S., ... & Ross, D. (1997). Benzene poisoning, a risk factor for hematological malignancy, is associated with the NQO1 609CT mutation and rapid fractional excretion of chlorzoxazone. *Cancer Research, 57*(14), 2839–2842.

Schrag, S. D., & Dixon, R. L. (1985). Occupational exposures associated with male reproductive dysfunction. *Annual Review of Pharmacology and Toxicology, 25*(1), 567–592.

Smart, C. R., Hendrick, R. E., Rutledge, J. H., & Smith, R. A. (1995). Benefit of mammography screening in women ages 40 to 49 years. Current evidence from randomized controlled trials. *Cancer, 75*(7), 1619–1626.

Taylor, L. H., Latham, S. M., & Woolhouse, M. E. (2001). Risk factors for human disease emergence.

Philosophical Transactions of the Royal Society London, 356, 983–989.

Thacker, S. B., Berkelman, R. L., & Stroup, D. F. (1989). The science of public health surveillance. *Journal of Public Health Policy*, 187–203.

Thacker, S. B., Parrish, R. G., & Trowbridge, F. L. (1987). A method for evaluating systems of epidemiological surveillance. *World Health Statistics Quarterly. Rapport Trimestriel de Statistiques Sanitaires Mondiales, 41*(1), 11–18.

Toxicological Profiles for Chlorinated Dibenzo-p-Dioxins. (1998, December 1). Retrieved from http://www.atsdr.cdc.gov/toxprofiles/tp104-c2 .pdf

US Preventive Services Task Force. (2009). Screening for breast cancer: US Preventive Services Task Force recommendation statement. *Annals of Internal Medicine, 151*(10), 716.

Wartenberg, D. (2001). Investigating disease clusters: Why, when and how? *Journal of the Royal Statistical Society: Series A (Statistics in Society), 164*(1), 13–22.

Winkelstein, W. Jr. (2004). Vignettes of the history of epidemiology: Three firsts of Janet Elizabeth Lane-Claypon. *American Journal of Epidemiology, 160*, 97–101.

World Health Organization (2013). Factsheet on the World Malaria Report 2013. Retrieved from: http://www.who.int/malaria/media/ world_malaria_report_2013/en/

Writing Group for the Women's Health Initiative Investigators (2002). Risks and benefits of estrogen plus progestin in healthy postmenopausal women: Principal results from the women's health initiative randomized control trial. *The Journal of American Medical Association 288*(3): 361.

Zimbardo, P. G. (2014) Stanford Prison Experiment. Retrieved from http://www.prisonexp.org/

GLOSSARY

Active immunity   Acquired through either previous exposure to the pathogen or immunization against the pathogen, which is usually through vaccination

Active surveillance   Survelliance initiated and completed by public health agency staff; labor and time intensive

Acute   Disease *acuteness* refers to rapid onset and/or short-course

Adjusted mortality rate   We adjust the mortality rates to remove the effect of age so the populations are similar in structure by indirect or direct standardization

Agents   Related to determinants in the definition of epidemiology

Airborne transmission   Disease caused by spores, dust particles, and through very small suspended droplets that enter the respiratory system

Analytical study   Generate hypotheses, examine associations, and attempt to find causal relations between exposures and outcomes

Assent   The process of asking a child to participate in a study

Association   Identifiable relationship between an exposure and an outcome

Attack rate   In food-borne outbreaks, *attack rate* is synonymous with incidence

Bias   Systematic error in the design or conduct of the study

Bias away from null   Occurs when the observed effect is stronger than the true effect

Bias toward the null   Occurs when the observed effect is weaker than the true effect

Binary   Dichotomous

Biostatistics   Statistical theory and methods, which are directly related to investigations in the biological and health sciences. In short, statistics is the art and science of collecting, organizing, describing, and analyzing data.

Blinded   Process of ensuring participant, physicians, and data analysts are unaware of which group participants are randomized into; way of reducing bias

Carriers   People who harbor the infectious agent but don't show any signs of infection

Case   Person who has been diagnosed as having a disease, disorder, injury, or condition

Case definition   Standard set of criteria to ensure consistency in diagnosis

Case fatality rate (CFR)   Number of deaths from a specific disease divided by the number of individuals in the population with the specific disease

Case report   Profile of a single indivudal

Case series   Group of case reports from individuals with the same disease

Case-control study   Study design that compares the exposures among individuals with a specific disease to those without the disease

Case-referent study   Synonym for case-control study

Categorical variable   Place individuals into one of several groups or categories with no numerical values

Causality   Relationship between exposure and outcome, where the exposure must temporally take place prior to outcome; be compatible with existing theory; show consistent results in replicated studies; risk of outcome is reduced by decline in exposure

Cause-specific mortality rate Evaluate death rates related to a particular disease or condition. It is calculated by looking at the number of deaths from a specific cause divided by the number of persons in a total population

Chronic disease Dieases and health conductions of continuous duration and must be managed appropriately; combination of components may be present

Cohort study Two or more groups of participants, with similar characteristcs except disease of interest, are followed over time simultaneously

Common source epidemic Occur when there is a pronounced clustering of cases of disease that occurs within a short time (i.e., within a few days or even hours) due to exposure of persons or animals to a common source of infection such as food or water

Common vehicle spread Spread of a disease through a common source—for example, through the air, water, food, or drugs

Concurrent study Group of participants is recruited in the present/current time and followed into the future to determine disease incidence; also referred to as prospective study

Confidence intervals Measure the range within which the true magnitude of effect lies with a certain degree of certainty

Confirmed case Following the information gathered, the individual is classified as *case confirmed* when all case definition criteria are met

Confounder variable that confuses the relationship between the exposure and disease

Confounding Distortion of an exposure-disease association caused by the association of another factor with both the diseases and exposure; variable that confuses the relationship between exposure and disease

Consistency causation Replication of *consistent* findings

Continuous source epidemic Occur when exposure to a source is prolonged over an extended period of time

Continuous variable Infinitive numerical values

Cross-sectional study Study design in which individual-level data is collected; both exposure and outcome data are collected at the same point in time

Crude mortality rate Number of deaths from all causes in a population divided by the number of persons in the total population

Cumulative incidence All the individuals identified as at risk at the beginning of the time period are followed for the specified period of time to determine how many develop the disease of interest

Death rate total number of deaths observed divided by the total amount of person-time; synnonym of mortality rate

Demographic transition Shift from high birth and death rates to much lower birth and death rates; occur as countries develop and grow economically

Dependent variable Response variable, is the health outcome or response being investigated

Descriptive statistics Describe important features and trends in the sample data

Descriptive studies Studies that observe the frequency of health-related states and provide a means of organizing, summarizing, and quantifying epidemiological data by person, pleace, and time

Detection bias Occurs if individuals with the exposure of interest are more likely to receive medical care and have the disease of interest identified

Determinant Any factor that brings about change in a health condition

Direct contact Person-to-person physical contact, such as touching with contaminated hands, skin-to-skin contact, kissing, direct droplet spread by sneezing or coughing, blood transfusion [via placenta included] or sexual intercourse

Direct standardization Method for adjustment requires that we select a standard population first

Discrete variable Count

Distribution Occurrence of diseases and other health outcomes varies in populations, with some subgroups of the populations more frequently affected than others

Dose-response relationship Effect of a toxic substance on the body is directly related to the strength of its dose; considered as evidence of a causal relationship

Ecological fallacy Making the assumption that an exposure is causal because it is prevalent in the same population that has a high prevalence of a certain health outcome

Ecological study   Use data aggregated at a population level, using groups of people as the unit of analysis

Effect modification   Occurs when the stratum-specific measures of association are not uniform; also known as interactions

Efficacy   Proportion of individuals in the control group who experience the outcome of interest (which is not desired), who could have been expected to have a favorable outcome had they been in the intervention group instead

Endemic   The constant presence of an agent or a health condition within a given geographic area or population

Environment   Surroundings and conditions external to the human or animal that cause or allow disease transmission

Epidemic   Occurrence of disease or other health-related events clearly in excess of what is normally expected; may only be one case

Epidemic curve   Graph that shows the distribution of cases of disease by time on onset of disease

Epidemiological transition   Shift in the patterns of disease and death from primarily acute, infectious disease to chronic, lifestyle-based diseases

Epidemiological triangle   Shows the interaction and interdependence of agent, host, environment, and time as used in the investigation of diseases and epidemics

Epidemiology   Sometimes referred to as 'population medicine'; study of the distribution and determinants of health and disease, injuries, diability, and mortality in populations

Equipoise   State of genuine uncertainty about the benefits or harsm that may result from each of the two or more regimens. A state of equipoise is an indication for a randomized controlled trial because there are no eithical concerns about one regime being better for a particular patient

Eradicated   Permanent reduction to zero of the worldwide incidence of infection caused by a specfic agent as a result of deliberate efforts; intervention measures are no longer needed

Etiology   Study of specific cause or combination of causes of disease

Exclusion bias   Establishing different eligibility rules for recruitment of the comparison groups in a study

Experimental study   Participants are placed into a treatment or control/placebo group and followed over time; include field, clinical, and community trial subtypes

Exposure   Pertains either to contact with a disease-causing factor or to the amount of the factor that impinges upon a group or individuals

External validity   Extent to which the results of the study are applicable to other populations

Fatality rate   Number of deaths in a specific population at a specific point in time divided by the total population in the same place and time

Fomite   Inanimate object that transmits diseases

Frequency matching   Controls selected for the study have a similar distribution of a matching variable among the cases

Herd immunity   Proportion of individuals in the population who are resistant to a particular disease

Hospital controls   Patients seeking care for conditions other than the disease of interest at hospitals or other health care facilities where cases are identified

Host   An organism, usually a human or an animal, who harbors a disease

Incidence   Number of new cases of a disease that occur during a specified time in a population at risk for the disease

Incidence density   To account for individuals not followed, the denominator in this calculation is the sum of the different times each individual was at risk, and it is often expressed in person-years

Incidence rate   To account for individuals not followed, the denominator in this calculation is the sum of the different times each individual was at risk, and it is often expressed in person-years

Independent variable   Explanatory variable, is the treatment, exposure, or risk factor(s) being studied

Index case   First disease case brought to the attention of the epidemiologist; not necessarily *primary case*

Indirect standardization   The indirect standardization process for adjustment is used when the age-specific death rates are unavailable

Indirect transmission   Pathogens or agents are transferred or carried by some intermediate item, organism, means, or process to a susceptible host, resulting in disease i.e. airbourne, vector borne or vehicle borne

Individual matching   Researcher matches each control to a particular case with respect to the matching variable(s)

Infectious disease   Caused by the entry and multiplication of microorganisms and parasites in the body of humans and animals

Inferential statistics   Used to investigate the research hypothesis about the source population, using information from the sample; allow us to conduct an investigation of a research hypothesis about a population of interest using information from a random sample of data from that population

Information bias   Error in the classification of participants with respect to disease or exposure status; most commonly referred to as misclassification and recall bias

Informed consent   Providing adequate information to each potential study participant so that he/she may make an informed decision about whether or not to participate

Innate immunity   Inborn barriers to disease and infection; physical barriers like intact skin, mucosal lining, cilia, and the cough-and-gag reflex; chemical barriers like acidity in the stomach, various enzymes, lipids, and interferons that create a hostile environment for agents seeking to invade

Institutional review board   Group at all institutions, including colleges and universities, responsible for ensure that the rights and welfare of all human research participants are adequate

Internal validity   Extent to which the results of a study accurately reflect the true situation in the study population

Latency period   Time between exposure and development of consequent disease

Lead-time bias   Occurs when screening detects a disease earlier in its course than if screening had not been performed

Length bias   Overestimation of survival duration among screening-detected cases, which is caused by the (relative) excess of slowly progressing cases

Life expectancy   The probable number of years a person will live after a given age

Life tables   Analyze how long a patient with a particular condition is likely to survive

Longitudinal study   Study design in which participants are followed forward in time; also known as a cohort study

Loss to follow up bias   Group of participants lost over the course of the study is systematically different from those who complete the study

Matching   The process of making the case and control groups as similar to each other as possible with respect to one or more variables

Meta-analysis   Combines the results from a number of studies using statistical methods to pool the results, providing a quantitative summary

Miasma   Scientific theory which blamed the spread of disease on bad air

Misclassification bias   Disease or exposure status of participants is categorized information

Mixed epidemic   Common source epidemic is followed by person-to-person contact, and the disease is spread as a propagated outbreak

Morbidity   Measurement of disease in a population

Mortality   Measurement of death in a population

Multifactorial etiology   Development of disease caused by many factors, including environmental exposures, social determinants (e.g., poverty), lack of access to care, etc.

Negative predictive value (NPV)   Quantifies the proportion of individuals with a negative screening test who truly do not have the disease

Nominal   Named categories

Null effect   Measure of association is the value that indicates no association between exposure and disease: a relative risk, or odds ratio equal to 1

Null hypothesis   States that there is no relationship between the independent and dependent variables

Odds ratio   Comparison of the occurrence of exposure among a group with the disease to those without disease

Ordinal   Ranked categories

p-value   Probability of obtaining a result at least as extreme as that observed in the study by chance alone. In other words, a p-value is a measure of compatibility between a hypothesis and the data

Pandemic Epidemic occurring worldwide or over a very wide area, crossing international boundaries, and usually affecting a large number of people

Participant Human subject being studied

Passive immunity Occurs less frequently and occurs when antibodies from other sources are given as postexposure prophylaxis for diseases like rabies and hepatitis

Passive surveillance Survelliance that requires that medical care providers—not public health practitioners—report notifiable diseases on a case-by-case basis to state and local health agencies

Period prevalence Prevalence of a disease in a population over a specified period of time (includes cases at the start of the period and any subsequent new cases)

Placebo Standard of care given to the group not receiving the experimental drug/therapy in a randomized controlled trial

Plausability causation Exposure/outcome relationship is logical and reasonable

Point prevalence Prevalence of a disease in a population at a single point in time

Point source epidemic Defined by the exposed developing the disease very quickly, often over one incubation period

Population All the inhabitants of a given area considered together

Population-based controls Random sample of individuals without the disease of interest who are selected from the same source population as the cases

Positive predictive value (PPV) Quantifies the proportion of individuals with a positive screening test who truly have the disease

Power The likelihood that the study will detect an association between a disease and exposure if an association actually exists

Precipitating factors Associated with the clear onset of a disease or illness

Predisposing factors Characteristics such as sex, age, educational status, marital status, work environment, previous or concurrent illness, and even attitudes toward using health services, or genetic traits that may result in an a weakened immune system or increased susceptibility to a disease-causing agent

Prevalence Count the proportion of the population with a disease risk factor or a disease at a particular point in time; calculated as the number of existing cases of disease at a specified time divided by the total population at the same time

Prevalence ratio Compares the prevalene of disease among those with the exposure to the prevalence of disease among those without the disease

Prevalence study Synomym for cross-sectional study

Primary case First disease case within a population

Primary prevention Prevent the occurrence of disease through personal and community efforts such as health education, improved nutrition, immunizations, sanitation, and infection control

Prognositic study Conducted in order to make a prediction of the future course of a disease; focus on both positive and negative outcomes including: death, complications, pain and suffering, quality of life, and remission

Propagated epidemic Arise from infections being transmitted from one infected person to another; transmission can be through direct or indirect routes

Proportion Fraction of the population is affected by the disease

Proportionate mortality rate (PMR) Proportion of the dead died from a particular cause. It is calculated by taking the number of deaths from a particular cause divided by the total number of deaths in the population

Prospective cohort study Group of participants is recruited in the present/current time and followed into the future to determine disease incidence; also referred to as concurrent study

Quantify To determine or calculate the amount of something (in epidemiology, we usually calculate the risk or odds of disease associated with some factor or exposure)

Quantitative variable Possess numerical value

Randomization Process by which all subjects have an equal probability of being assigned to either the intervention or control group

Randomized control trial Clinical control trials in experimental studies

Rate   Indicates how fast the disease is occurring in a population; a measure of the frequency with which an event occurs in a defined population in a defined time

Ratio   Used to compare the occurrence of a variable in two different groups, neither of which is included in the other

Recall bias   Cases remember and report risk factors in a different way than the controls

Reference population   Also referred to as *standard population*; reflects the concept that death occurring at a younger age results in greater loss of future productivity than would death at a later age

Reinforcing factors   Repeated exposures, environmental conditions, or work-related activities or behaviors that aggravate or perpetuate an established disease or an injury

Relative risk   Compares the risk of disease in the group with the exposure to the risk of disease in a group without the exposure; also known as the risk ratio

Research hypothesis   States that the relationship we expect to find between the independent and dependent variables

Reservoir   Normal habitat in which infectious agents live, grow and multiply

Response bias   Systematic error due to differences in the characteristics between participants

Retrospective cohort study   Group of participants be assembled in the past, and disease incidence is calculated in the present

Risk factors/determinants   Any factor that brings about change in a health condition

Sampling bias   A non-random sampling strategy

Secondary prevention   Prevention attempts to reduce the progress of disease through early detection and prompt interventions; i.e. screening tests

Selection bias   Results in differences in the characteristics of those who are selected to participate and those who are not; or the characteristics of groups with the study: specifically, the association between the exposure and disease differs for those selected into the study and those who are not

Sensitivity   Ability of a screening test to correctly identify individuals with the disease

Serial transfer   Transmission of disease from human to human, human to animal to human, or human to environment to human in a sequence (measles, STDs, AIDS).

Source population   The population the study is designed to make inferences about

Specificity   Ability of a screening test to correctly identify individuals without the disease

Standard population   Also referred to as *reference population*; reflects the concept that death occurring at a younger age results in greater loss of future productivity than would death at a later age

Stratification   Separate into different groups

Study design   The blueprint that allows for an assessment of events for statistical inference concerning relationships between exposures and diseases

Survelliance   Ongoing systematic collection, analysis, and interpretation of health data essential to planning, implementation, and evaluation of public health practice, as well as the timely dissemination of these data to those who need to know

Susceptible   Individual is vulnerable to a disease because he or she has no resistance or immunity to the disease

Suspected case   Individual who has all of the signs and symptoms of a disease or condition, but has not yet been diagnosed via laboratory or other definitive testing methods

Switchover   Observed and true effect are on either side of the null value

Symptomatic cases   Individuals who have apparent signs of the infection

Syndromic surveillance   Uses symptom information to alert public health officials to a potential problem

Systematic review   Uses defined strategies to find, analyze, and synthesize all relevant studies on a particular topic

Target-organ specificity   Toxic substances do not affect all organs to the same extent

Temporal relationship   Exposure must have occurred before the disease developed

Tertiary prevention   Focuses on reducing impairment and minimizing the suffering of patients with disease through rehabilitation and palliative care

Time   Time accounts for incubation periods (the time between the host first encountering the pathogen and when signs and symptoms first appear), life expectancy of the host or the pathogen, and duration of the course of the illness or condition

Toxicology   Branch of science concerned with the study of chemicals and their effects on the human body

Transmission   Any mechanism by which an infectious agent is spread, and is essentially how an infectious agent bridges the gap between portals

Variolation   Method of immunizing patients against smallpox by infecting them with the substance from the pustules of patients with the disease

Vector borne   Disease transmitted through a live intermediatry, such as an insect or animal

Vehicle borne   Disease spread through inanimate intermediates, including food and water, clothes, bedding, cooking utensils, and medical equipment

Volunteer bias   Occurs because people who are healthier, health-conscious, or have medical insurance are more likely to be screened than those who are less healthy or lack insurance

Years of life lost (YLL)   Calculated by subtracting a person's age at death from the the standard life expectancy for the genral population of interest

Years of potential life lost (YPLL)   Also referred to as *years of life lost* (YLL)

Zoonoses   Any disease or infection that is naturally transmissible from vertebrate animals to humans

α-level   Probability of observing an association between the disease and exposure of interest exists, when the truth is that association does not exist

β-level   Probability of failing to observe an association between the disease and exposure of interests exists, when the truth is that association does exist

INDEX

CREDITS

Figure 1.1: Source: http://commons.wikimedia.org/wiki/File:Wash_your_hands_poster_CDC_-_Wellbee.jpg Copyright in the Public Domain.

Figure 1.2: Source: http://commons.wikimedia.org/wiki/File:PHIL_4471_lores.jpg. Copyright in the Public Domain.

Figure 1.3: Copyright © ChiefHira (CC BY-SA 3.0) at http://commons.wikimedia.org/wiki/File:Damage_of_Tsunami_in_Kesennuma.JPG.

Figure 2.1: Source: http://commons.wikimedia.org/wiki/File%3ASotarl%C3%A4rlingen_Viktor_Norin%2C_Stockholm._Kabinettsbild%2C_1880-tal_-_Nordiska_Museet_-_NMA.0041502.jpg.

Figure 2.2: Source: http://commons.wikimedia.org/wiki/File%3AEdward_Jenner_vaccinating_his_young_child.png. Copyright in the Public Domain.

Figure 2.3: Source: http://en.wikipedia.org/wiki/File:John_Snow.jpg. Copyright in the Public Domain.

Figure 2.4: Source: http://commons.wikimedia.org/wiki/File:Snow-cholera-map.jpg. Copyright in the Public Domain.

Figure 2.5: Source: http://commons.wikimedia.org/wiki/File:Cholera_Epidemic_poster_New_York_City.jpg. Copyright in the Public Domain.

Figure 2.6: Source: http://commons.wikimedia.org/wiki/File:Louis_Pasteur_experiment.jpg. Copyright in the Public Domain.

CPSIA information can be obtained
at www.ICGtesting.com
Printed in the USA
LVOW02s1214060917

547691LV00003B/4/P